THE MOUNTING AND LAMINATING HANDBOOK

SECOND EDITION

Other Books Written or Illustrated by Chris A. Paschke

The Mounting And Laminating Handbook, First edition, 1997
Creative Mounting, Wrapping And Laminating, Designs Ink PUBLISHING, *1999*

Feng Shui: The Art of Living, Peter Pauper Press, 2000
The Feng Shui Journal, Peter Pauper Press, 2001
Feng Shui for the Home, Peter Pauper Press, 2002

THE MOUNTING AND LAMINATING HANDBOOK
Second Edition
By Chris A. Paschke, CPF GCF

Illustrations: Chris A. Paschke, CPF GCF
Editor: C. A. Paschke
Paste-up and Proofreading: C. A. Paschke, T. A. Irvin
Computer Formatting: Thaer Anthony Paschke Irvin
Published by
Designs Ink Publishing
Tehachapi, CA USA

Copyright © 2002, 1997 by Chris A. Paschke, CPF GCF
All Rights Reserved.
Translation or reproduction of any portion of this book,
beyond the International Copyright Act, without the
express written consent of the publisher is unlawful.
All inquires should be addressed to:
Permissions Department, Designs Ink Publishing,
183-785 Tucker Road, Suite G, Tehachapi, CA 1.661.821.2188
www.designsinkart.com

Printed in the United States of America
10 9 8 7 6 5 4 3 2

ISBN 0-9657625-2-1
Library of Congress Catalogue Control Number 2001127223

The purpose of this handbook is to educate and enlighten, and was written with the understanding that the publisher and author are not responsible for the outcome or results of any projects. Mounting and laminating methods described herein are based on standard practices routinely used in the professional workplace. It is not the purpose of this handbook to establish rigid procedures for handling fine art and photographs, but rather to establish a constructive set of guidelines for individual project decisions. Every effort has been made to make this handbook as complete, accurate and up-to-date as possible. However, there may be mistakes, both typographical and in content.

It should be understood that preservation and conservation methods do not involve altering the artwork in any way. The mounting and laminating techniques are in no way conservation procedures. It is the exclusive responsibility of each individual professional picture framer to select and execute the proper procedure for each individual project. The publisher and author shall have neither liability nor responsibility to any person or entity with respect to any loss or damage caused, or alleged to be caused, directly or indirectly by the information contained herein. If you do not wish to be bound by the above, you may return this book to the publisher for a full refund.

CHRIS A. PASCHKE, CPF GCF

The Mounting And Laminating Handbook

SECOND EDITION

DESIGNS
INK
PUBLISHING

Dedicated to

Dad,
for showing me that
perfectionism is a virtue, not a vice.

ABOUT THE AUTHOR

Chris A. Paschke, CPF GCF is a second generation picture framer, educator, demonstrator, designer, industry consultant and specialist in mounting and laminating. Having originally received *Certified Picture Framer (CPF)* status from the Professional Picture Framers Association (PPFA) in 1986, she was awarded *Guild Commended Framer (GCF)* status from the Fine Arts Trade Guild (FATG) in 1997 from the UK. Both acknowledgements represent an award of excellence in craftsmanship and service in the picture framing industry.

She is a long time supporter of PPFA as unfaltering member, chapter leader and competition judge, currently serving on *PMA/PPFA Master Educator, Awards and Recognitions* and *ReCertification Committees*. She has written the monthly column "Mastering Mounting" as Technical Mounting Editor for <u>Picture Framing Magazine</u> since 1991, and is an ever present faculty member at industry trade shows. Having a degree in Creative Arts and Design, her passion in education is in design and creative applications of traditional and standard techniques.

Having published the first edition of *The Mounting And Laminating Handbook* in 1997, she proudly releases this, her second edition, as an expanded and updated version. She released her second book *Creative Mounting, Wrapping and Laminating* in January 1999, and continues to work on her third book *The Principles of Framing Design*, which hopes to see publication in 2003.

Acknowledgments

Prior to writing my first book, I never understood what an overwhelming task writing a book could be. Winston Churchill once said that "...writing a book goes through five phases. In phase one, it is a novelty or a toy. But by phase five, it becomes a tyrant ruling your life. And just when you are about to be reconciled to your servitude, you kill the monster and fling it to the public."

This being my revised second edition was a little easier than the first, but the torment still exists with every new endeavor. All the people that helped me with the first release of this book in 1997, especially Leslie Denton, Greg Fremstad, Brian Wolf, and Greg Perkins are still close to my heart, and the text that was there then still remains alongside the new information.

In this edition I wish to thank the following individuals and companies who have been supportive through the sharing of their technical knowledge, product information, comments, suggestions and advice for all of the new information: Don Dressler, 3M, Crescent Cardboard, David Kopperl, Drytac Canada, Epson, FrameTek, Larrie Deardurff, Hunt Corporation, Ilford, Nielsen Bainbridge, Polaroid Corporation, Print Mount, Neschen, Pilar Martinez at Rochester Institute of Technology, Roy Rowlands GCF, Seal Graphics Americas, Tullis Russell Hot Press, Mark McCormick-Goodhart at Wilhelm Institute, and Xerox. And a special thank you to the numerous others that have submitted information upon request and generally gone the extra mile to help me during this endeavor...you know who you are.

Thank you all.

Contents

List of Tips
List of Diagrams

Preface

PART ONE   MOUNTING

1   MOUNTING BASICS   5
       What is Mounting?
       When to Mount and When NOT To
       Disclaimers   6
       Statistics
       How to Sell Mounting   7
       Pricing   8
       Heat Mounting: The Whole Profit Picture
       Investment Payoff   10

2   UNDERSTANDING MATERIALS   11
       Adhesives Types
       Wet Glues   12
       Sprays   12
       Pressure-Sensitives   12
       Heat-activated
           Type of Bond   13
           Physical Composition
           Degree of Porosity   14
           Acidity Level or pH   15
       Preadhesived Boards   15
       Substrates   16
           Traditional Basics
           Mat Boards   17
           Foam Boards
           Heavy Duty Foam Boards   18
           Corrugated Backing Boards
           Archival Photo Boards   19
           Fabrics

	Release Materials	20
	Clear Films	
	Double-Sided	
	Single-Sided	21
	Boards	22
	Solvents	23
	Removing Adhesives	
	Tapes	
	Wet Glues and Sprays	
	Dry Mount Tissues and Films	
	Heat-activated Tissue from Prints	24
	Pure Film Adhesives	
	Permanent Tissue from Photos	
3	**SETUP AND TROUBLESHOOTING**	25
	Clean Area...Clean Process	
	Workstation	25
	Lighting	
	Storage	26
	Placement/Location	27
	Mechanical Press Maintenance	28
	Daily Routine	
	Cleaning	29
	Vacuum Maintenance	30
	Adjustments	
	Daily Routine	
	Cleaning	31
4	**ELEMENTS OF MOUNTING...TTPM**	32
	Time	32
	Mechanical Press	
	Vacuum Press	
	Temperature	33
	Wet and Spray	
	Dry Mounting	
	Time/Temperature Ratios	
	Pressure	34
	Wet and Spray	
	Pressure-Sensitive	
	Dry Mount: Variations	
	Moisture	35
	Wet and Spray	
	Pressure-Sensitive	
	Dry Mount: Mechanical	
	Dry Mount: Vacuum	

5	**WET MOUNTING**		37
	Basic Principles		
	Tips and Techniques		38
	APPLICATIONS AND PROCEDURES		39
	Standard Wet Mounting		
	Brush Applications		
	Countermounting		40
	Weighting		
	Oversized Poster Mounting		41
	Floated Art/Japanese Papercuts		
	Encapsulation		42
6	**SPRAY MOUNTING**		43
	Basic Principles		
	Tips and Techniques		44
	Adhesive Types		46
	APPLICATIONS AND PROCEDURES		47
	Standard Spray Mounting		
	Japanese Papercuts		48
7	**PRESSURE-SENSITIVE MOUNTING**		49
	Basic Principles		
	Tips and Techniques		51
	APPLICATIONS AND PROCEDURES		
	Pressure-Sensitive Films		52
	Standard Film Application		
	Self-Shaping		53
	Ghosting Newsprint		
	Color Tinting		
	Float, Plain or Flush Mounting		54
	Pressure-Sensitive Boards		55
	Standard Board Application		
	RC Photographs		57
	Ilfochrome Classic (Cibachrome)		
	Brass Rubbings/Digitals		57
8	**COLD MOUNTING**		58
	Basic Principles		
	Cold Laminating		59
	Mounting Polyester Encapsulates		
	Vacuum Frame Basics		60
	Flattening Artwork		
	Heat-sensitive Items		61
	Thermography		
	Digital Imaging		62
	Ilfochrome Classics		66
	Static Mounting		68

9	**DRY MOUNTING**	69
	Basic Principles	
	Tips and Techniques	70
	Weighting	
	Tacking Basics	71
	Surface Tacking	
	Z-Method Tacking	72
	Multiple Bites Tacking	73
	Presses	
	Mechanical (Softbed) Presses	75
	Checking Pressure	
	Pressure Adjustments	76
	Shimming Variations	
	Predrying for Moisture Control	
	Hardbed Presses	77
	Hot Vacuum	78
	Review of Hints and Reminders	79
10	**DRY MOUNTING APPLICATIONS**	81
	Basic Techniques	
	Adhesive Trimmed to Size	82
	Oversized Adhesive	83
	Premounted Adhesive	84
	Montage or Multiple Mounts	85
	Plain or Float Mounting	86
	Flush Mounts	87
	Multiple Bites	88
	Translucent Materials	
	Rice Papers and One-Sided Text	89
	Newspapers and Two-Sided Text	90
	Silks and Sheer Fabrics	91
	Color Tinting	
	Opaque Materials	
	Fabric Wrapping	92
	Wrinkled Paper Technique	94
	One-step Shadow Boxes	96
	Jigsaw Puzzles	98
	Brass Rubbings/Papyrus/Vellum	99
	Photographs	
	RC Photos	100
	Oversized Photographs: Mechanical Press	101
	Oversized Photographs: Vacuum Press	102
	Fibre-base (Silver Gelatin)	104
	Ilfochrome Classics (Cibachrome)	105
	Heat-sensitives	105

PART TWO   LAMINATING

11   LAMINATING — 109
    What is Laminating?
    Surface Laminating vs. Encapsulating
    Marketing Potential — 110
    Polyester Films — 111
        Two-Sided Encapsulation
        Roller Laminators
    Vinyl Films — 112
        Surface Lamination
    Film Overview — 113
    Additional Tools and Materials — 114
        Overlay Foams
        Foam Plastic
        Foam Troubleshooting
        Perforators

12   LAMINATING APPLICATIONS — 115
    Basic Techniques
        Standard Alignment, Preparation and Set-up — 115
        Surface Laminating Breathables — 117
        One-step Mounting/Laminating — 118
        Handling and Laminating Oversized Art — 119
    Photographs
        Perforating for Nonbreathables — 120
        Surface Laminating Photos — 121
        Two-Step Process — 122
    Canvas Transferring — 123
        Copyright
        Equipment and Material Variations
        RC Photographs (Chromogenic) to Canvas — 124
            Mechanical Press
            Vacuum Press — 125
        Fibre-base Photos (Silver Gelatin) — 128
        Untreated Canvas/Watercolor Papers — 129
        Poster Prints — 130
        Encapsulation — 132

PART THREE   APPENDIX

US/UK Terminology Equivalents / F°-C° Conversion Chart	135
Suggested Mounting Methods	136
Photographic Recommendations	138
Flattening Photos	140
Selecting a Press/Pros and Cons	142
Adhesives	144
Digital Technologies	145
Copier Heat and Laminate Tolerances	150

BIBLIOGRAPHY	153
INDEX	154

LIST OF ➤ *TIPS*

Air Bubble	86
Air Bubbles and Their Removal	102
Alcohol Cleaning	42
Alternate Heating for Removal	23
Archivally-Named Adhesives	104
Brush Strokes for Canvas Texturing	131
Computer Art	80
Controlling Steam	73
Creeping P-S Adhesive	54
Digital Images	42
Digital Photographs	87
Encapsulating Prints	131
Foam Alternative	56
Giclees	36
Heating Sprays for Removal	46
High Gloss Finishes on RC Photos	102
Identifying RC Photos	101
Ironing Bevels and Tabs	94
Mounting Digital Images	65
Paper Moistening for Canvas Contour	128
PEC 12	19
Pollutants	31
Release Boards and Orange Peel	101
Reproductions	90
Rigid Board Stiffener	125
Saturation of Adhesives	91
Self-Shaping Pressure-Sensitive Films	47
Solvent Sensitive Inks	22
Suction Sealing	86
Suffocation of Multiple Nonporous Layers	121
Temperature Variations for Canvas Transfers	123
Thermographic Paper Darkening	94
Wrinkle Removal	46

LIST OF DIAGRAMS

Chapter 1	Sales samplers	7
	Mounting and laminating pie chart	8
	Whole profit picture	10
Chapter 2	Tissue-core vs. film adhesives	13
	Nonporous art with porous tissue	14
	Nonporous art with nonporous tissue adhesive	14
	Porous art with porous tissue or film adhesive	14
	Release boards top only and release sandwich	21
Chapter 3	Adhesive roll dispensing rack, front and side views	26
	Preparation table, press, and cooling table	28
	Unhinging a mechanical press	29
	Release wrap around sponge pad	29
Chapter 5	TTPM for wet mounting	37
	Countermounting	40
Chapter 6	TTPM for spray mounting	43
	Spray booth	45
	Spray accordion support	43
Chapter 7	TTPM for pressure-sensitive mounting	49
	Application of p-s film with two-sided release liners	52
	P-S board application	55
Chapter 8	TTPM for cold mounting	58
	Static mounting Ilfochrome Classics	68
Chapter 9	TTPM for dry mounting	69
	Weighting	70
	Tacking positions	71
	Z-method tacking	72
	Multiple bites in a mechanical press	73
	Scored 45-degree pattern for mechanical press	75
	Adjusting for pressure	76
	Suction sealed edges	80
Chapter 10	Standard mounting package	81
	Premounting adhesive	84
	Premounting multiples for montage	85
	Plain or float mounting	86
	Flush mounting square, bevel, and reverse-bevel cut edges	87
	Two-sided text on colored substrate, or ghosting	90
	Fabric wrapped mats	93
	Wrinkled paper wraps	94
	Paper grain	95
	One-step shadow box	97
	Trapped air beneath large vacuum mounted RC photo	102
Chapter 11	TTPM for polyester laminate films	111
	TTPM for vinyl laminate films	112
	Polyester/Vinyl laminate overview	113
	Perforators	114
Chapter 12	One-step mounting and laminating	118
	Perforator and application of holes into film	120
	RC photos transferred to canvas	124
	Untreated canvas and laminating film	129
	Poster prints transferred to canvas	130

Preface

When I began framing with my father in 1970, it was perfectly acceptable to glue images to corrugated cardboard and use masking tape to hold artwork into mat windows. Today there are more sophisticated methods designed to better protect and enhance valued artwork. This handbook is the love child of many years of acid burn, mounting damage, trial and error...and was written in the hopes new framers will not have to blaze the same trails my father and I did so many years ago.

After two printings of the first edition of this book and the emergence of digitals onto the art scene, my first edition required additional information and grew into a proud second edition, boasting sixteen additional pages and information about digitals. This revised second edition has new explanations, more tips, test results, and mounting suggestions for images and emerging art in the 21st century. Five years ago the CMC (computer mat cutter) was the newest piece of frame shop equipment. Ten years ago it was the underpinner, fifteen years ago the heat vacuum press, today the 21st century frame shop will most likely consider investment in a cold roller laminator to accommodate the heat sensitivities of the digital world.

This book is laid out and structured to teach procedures rather than specific projects. In this way, you will always be able to apply the learned technique to each individual project, as needed. You will then be better armed to answer your own questions and truly understand products, materials, AND applications. Please use this handbook aggressively. Make notes in it, dog-ear the pages and work it to death, it's your textbook, your handbook. The more worn-torn it gets the better it will help you and the better I will have done my job. Good luck all.

Keep it close to your heart for me, and close to your mounting press for you.

Chris

PART ONE

Mounting

Mounting Basics

WHAT IS MOUNTING
Mounting is the action of affixing a paper image, photograph, or fabric to a stiffer support, backing, or substrate. It may be attached in a number of ways from conservation hinges to the more permanent technical procedures utilizing wet, spray, pressure-sensitive, cold vacuum or heat mounting methods.

Expansion and contraction of individual paper layers during normal humidity and temperature changes may be controlled and/or eliminated by mounting. The concepts of mounting in this book involve the physical attachment of artwork to a substrate to flatten and prevent it from bubbling, cockling, or warping. These are not to be considered conservation methods.

WHEN TO MOUNT AND WHEN NOT TO MOUNT
Knowing when NOT to mount an item is equally as important as HOW to mount it successfully. Quite often, it is not a question of *can it be safely mounted*, but rather *should it be mounted* at all. The list of what not to mount varies, but generally includes signed limited editions, vellum, parchment, university certificates, documents with verso notations, Ilfochrome Classics (a.k.a. Cibachromes), original art of any kind and irreplaceable heirlooms. Though these items may mount without harm or damage to them, they will no longer be truly reversible, or able to return to their original state.

Heat mounting limitations include thermographics such as tickets, embossed plastic lettering, sensitive digital prints, some inkjet and laser prints, and color copies. Some of these items turn black or melt when subjected to heat, while others we may simply not know about. If there is any question concerning mounting sensitivity or value of the artwork, **do not mount it**. Use only conservation/preservation methods when mounting.

Always preserve the integrity of the profession and the art by advising the customer to do everything possible to **enhance and protect the art**. Mounting the art might not always be in the best interest of protecting it. An original watercolor may look better mounted flat, but it is not the correct procedure for protecting it. Only the artist has the right to mount their originals, never the customer or framer.

Disposable or decorative art includes any inexpensive poster, offset print, or digital designed to be used as short term wall decor. Posters, RC photographs and fabrics are all readily mountable and at least 90% of all decorative art being framed should be mounted. There is no perfect answer to the mounting question and no single solution to any given mounting project. Though mounting needs may be met by a single process there will generally be more than one solution to every problem, and each will often designate which technique is best suited to it.

MOUNTING DISCLAIMERS
Most customers assume their art is covered by framer's insurance whether it be cut, drowned, or fried. Always be totally clear with the customer about liabilities, especially when mounting is part of the project. It is best to indicate any potential problems, sensitivities, or risks and discuss liability limitations at the time of sale. Customers need to learn that framers handle and mount in the best way possible, and that mounting often involves educated guesses in terms of the heat and moisture sensitivity of the art. These guesses are not truly confirmed until the completion of a project. There is an element of risk involved in mounting any project.

Mounting procedures should always be predictable and routine. Moreover, any project slated for mounting **must have the ability to be replaced**. Unusual requests or unfamiliar materials should be carefully thought through before mounting. Never mount anything common sense warns will not tolerate heat, moisture, or both. There are always conservation alternatives, such as Japanese hinges, Mylar corners, or edge strips.

DIGITALS
With the advent of digitals comes the unknown. Without confirmation of the printing technology, inks and substrate used, it is difficult to determine which mounting method is best for any given image. Digital inks may be heat or moisture sensitive, and some substrates resist absorption, allowing inks to more easily flake off. Serious care is required when mounting, and traditional equipment or methods may require testing prior to implementation.

During the design process it is a good idea to request duplicate images when framing a digital, one to test for mounting tolerances and then one to frame. This is not to say framers are negligent or tend to damage projects, but rather it is the sign of the times. Just as in the past it was asked what color the carpet and period of furniture...today we need two digital prints.

STATISTICS

If nondigital open edition graphic prints (posters) are part of the framing inventory, then 90% of what is sold should be mounted before it leaves the store. This is tapping into the potential profits from press use, whether cold frame, hot vacuum or mechanical press. Whether as a complete glassed and framed project or a shrink-wrapped poster, mounting ensures sales and encourages even inexpensive items to be framed.

Currently, statistics show only about 25% of all frame shops own heat mounting equipment. This illustrates the volume of alternative mounting methods being used today and marks the potential of market sales. Heat systems open sales up to laminating and creative applications as well as mounting, while roller laminators ensure successful pressure-sensitive mounting and laminating for digitals.

HOW TO SELL MOUNTING

The beauty of press mounting includes the speed with which prints may be mounted, ease surrounding the process, and long term permanence of a professional presentation. If it looks good the customer remains happy, but if it bubbles or cockles the customer may never return. There are a few basic techniques for selling mounting. Using open ended questions; having answers prepared to meet potential objections; and having physical samples to visually better explain the mounting process itself, all promote closing "the sale."

Always sell by asking open-ended questions, the kind that encourage additional discussion..."Tell me about the colors of your room." Closed questions elicit a simple yes or no answer only. Never ask, "Do you want to have this mounted?" You might hear "No." When it comes to the recommended procedures, say "This price includes the frame, mat package as we designed it, and glazing for a total of $_. Of course that includes all mounting and fitting charges." One approach to overcoming any mounting objection would be to offer the option of mounting at a later date. Just point out the additional charges, for both unfitting and refitting, that will need to be included besides the mounting when that is done.

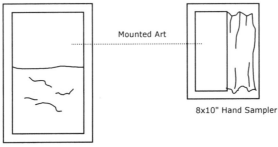

24x36" Framed Poster   8x10" Hand Sampler

As a final objection comeback, a sampler of half mounted, half cockled paper is worth 1000 words to help make the mounting sale. A large 24x36" thin, inexpensive, dark poster is perfect. Strategically place the framed sample to optimize visual glare from windows when standing in the customer's position at the design counter, with a sign promoting the benefits of mounting. A smaller 8x10" sample should be kept behind the counter for closer examination. Never hesitate to educate the consumer. It is your job.

PRICING

Different mounting procedures price very similarly when hard costs are added to actual labor time. This means that if materials cost less, but the process takes more time to produce, such as wet mounting, it nearly equals the expenses of the higher priced dry mounting costs which takes less time to complete. By calculating united inch prices of four basic sizes (8x10, 16x20, 24x30, 30x40), then averaging them together, wet mounting actually comes in at the highest suggested retail price.

Manufacturers favor square foot calculations for setting prices over the more traditional united inch calculations when it comes to mounting. Suggested pricing charts are often calculated on a united inch format, but using that formula may not totally cover actual hard cost and overhead. Verify adhesive expenses to check mounting charges. Do not forget the difference in preparation, execution and cleanup also adds to labor overhead when calculating wet and spray mounting applications. Plus, always consider what the market will bear when setting prices.

HEAT MOUNTING: THE WHOLE PROFIT PICTURE

When considering the capital investment of a large piece of mounting equipment, common concerns and questions often include space requirements, service after the sale, education, payoff, and profit potential. In order to determine which system is the best, anticipate the number and types of mountings per week; the average size of mountings; how many different individuals will be using the system; and whether the full profit potential of owning a mounting system has been considered.

Mounting is only 1/2 of any heat press potential and 1/3 of the total profit available. Most framers do not run their presses all day every day, but only when projects are due. If, however, a targeted market is mounting and there is a press operator working the machine every minute of the day, maximum profits are indeed being felt.

Integration of laminating into frame shop services will require additional marketing to ensure growth. Yellow pages advertising, mailers, and networking all provide sources for telling the world of new services available.

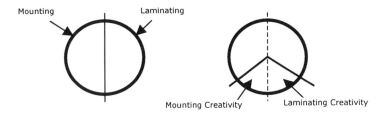

Only half of a heat press's potential use is mounting . . . the other half is laminating.

Potential sales for surface laminating will be found wherever glass may not be used. Any place dealing with children, such as day-care centers, preschools, nurseries and Pediatricians' offices would be perfect. The same is true with the elderly at nursing homes and hospitals. Detention facilities, jails, and restaurants are all places where safety is an issue or do not allow glass. Engineering and architectural firms, city, state, and federal offices may need laminating for presentation to groups or clients. Real estate agencies obviously need maps for both pinning home locations and for drawing on with washable markers. Also consider sporting goods stores, hunting, fishing, hiking and camping information centers or recreational areas for mounting and laminating needs. The possibilities are endless.

Once mounting and laminating are both offered there is yet another profit making segment which still may be untapped. This is the creative portion of mechanical and hot vacuum press usage, where the most fun and greatest profits lie. Tiered matting with colored core boards and art papers; one-step shadow boxes; wrapped and embossed mats all fall into mounting creativity...while resurfacing, refinishing and retexturing films; leather-look; contempo panels; faux glass etching; mirror designing; embossing; and canvas transferring of both photos and prints overlap into laminating creativity. These creative techniques are taught in workshops, at open houses, and trade shows around the country. They are also routinely written about in trade magazines with techniques available in *Creative Mounting, Wrapping and Laminating*, published in 1999.

As long as any press (hot or cold) is used continually for mounting during the course of each business day it is being used 100%. If it is also used for both mounting and laminating all day it is also used to its maximum 100%. If however, it sits idle some of the day AND no laminating OR creative applications are employed, then only 1/3 of the total available profits are being realized. See chart next page.

INVESTMENT PAYOFF

Anytime investment in a major piece of equipment is being considered, research should be done to establish an approximate time required to payoff or recoup the actual investment. This needs to be done before any incoming dollars become profits. If considering **mounting** only when calculating equipment payoff, it will take much longer to recoup investment dollars than if including laminating in the equation. The entire approach to successful profit dollars comes with dry mounting that also actively includes **laminating** and **creative applications**.

For every investment figure of $2000...

MOUNTING 20pcs week @ $ 10. = $200. divided into $2000. = 10 week payoff
added to
LAMINATING 10pcs week @ $ 20. = $400. divided into $2000. = 5 week payoff
added to
CREATIVITY 2pcs week @ $100. = $600. divided into $2000. = **3½ week payoff**

The above estimate illustrates gross sales potential, not including overhead and electricity. To calculate net profit, deduct overhead.

2

Understanding Materials

ADHESIVE TYPES
An adhesive is a substance capable of holding two surfaces together in a strong bond. This close bond is between the surface molecules of the material being adhered and the substrate. The closer the two surfaces fit together, the stronger the bond. Therefore a thin adhesive bond is often stronger than a thick one. Two main adhesive types are *natural,* substances coming from animal and vegetable sources; and *synthetic,* being compounded from simple chemicals, many of which are polymers. Vegetable glues come from starches and dextrins extracted from corn, potatoes, rice or wheat. Natural gums, although vegetable in nature, are often blended with synthetic rubbers to create adhesives often used in pressure-sensitive cellophane and masking tapes.

Synthetic adhesives fall into two categories: *thermoplastic* and *thermosetting.* Thermoplastic adhesives can be resoftened any number of times by reapplying heat and will once again adhere and bond when cooled. They are also soluble in selected solvents. Natural adhesives are predominantly thermoplastic, the most widely used being vinyl resin adhesives, more commonly known polyvinyl acetate (PVA) or white glue.

Thermosetting adhesives undergo an irreversible chemical change when they harden, the result of a catalyst. Once hard, they do not melt or resoften when heated and are considered insoluble in common solvents. Thermosetting adhesives include epoxies, polyesters, and urethanes such as used with fiberglass. These adhesives adhere well to most materials, porous and nonporous.

In framing, the adhesives used in mounting artwork to substrates include wet, spray, pressure-sensitive and heat-activated materials. The natural or synthetic base of an adhesive will categorize and in turn determine its appropriateness for any given type of mounting. Some adapt best to mass production, where speed and permanence is desired, while others allow for specialized attention during a slower mounting process.

WET GLUES

Thermoplastic in nature these water-based adhesives include, vegetable starch and PVA, and may be found as both wet and spray glues. Nonremovable polyvinyl acetate glues (PVA) are water-soluble when wet, but like any acrylic are quite permanent once dry. Wet glues are bottled liquids that are applied to a substrate with a brush, rubber roller, or airbrush. They require the pressure of a weight or cold vacuum frame to create the initial bond, which will hold for an indefinite period. Vegetable starch glues remain removable with distilled water even after long-term mounting.

SPRAYS

These are air-drying adhesives that are generally thermoplastics. They convert to a solid state by evaporation of the solvent, and bond by mechanical adhesion (the adhesive strength given by interlocking molecules) and mutual attraction. Rubber or acrylic-based solvent adhesives are found both as pressure-sensitives and spray glues. Sprays are packaged in aerosol cans and require sufficient pressure to create an adequate bond. Since partial adhesive saturation is required to ensure the best bond, when mounting some high gloss papers, RC photos and polyester fabrics, a truly permanent bond is difficult to achieve.

PRESSURE-SENSITIVES

These are predominantly thermoplastics, and require no moisture or heat prior to application. They are found as adhesive or adhesive/board combinations with release liners applied to protect them from bonding until desired to do so. Application of water or solvent before application activates some tapes. Pressure-sensitive adhesives are available in rolls, precut sheets and premounted to various substrates. The adhesive has a release paper backing, is tacky to the touch, and best results are derived from direct pressure of a roller press or squeegee applicator.

HEAT-ACTIVATED ADHESIVES

Dry mounting adhesives are easiest to understand when broken into specific categories. By taking the time to analyze the various available tissues, you will be better prepared to select the proper adhesive to fit your needs. All heat-activated adhesives may be placed into specific categories of:

- Type of bond (permanent or removable)
- Physical composition (tissue-core or film)
- Degree of porosity (breathable or nonbreathable)
- Acidity level (tissue pH)

TYPE OF BOND (PERMANENT OR REMOVABLE)
One of the keys to successful dry mounting is remembering where the bonding actually occurs. A permanent adhesive bonds within the press. All layers of the mounting package (top release material, art, adhesive, substrate and bottom release material) must reach the required bonding temperature and remain there during the required time allotment to set the adhesives. When removed they will be bonded.

A removable adhesive bonds once removed from the press as it cools under a weight. It becomes removable through the reapplication of heat, which reactivates the adhesive, making the art separable from the mounting substrate. All mounted items should be placed under a weight when removed from the press, regardless of whether permanent or removable, to expedite the cooling and help reflatten bowed substrates. It remains the suggested procedure regardless of where the adhesive bond occurs.

PHYSICAL COMPOSITION (TISSUE OR FILM)
Dry mount adhesives are available in both roll and precut sheets, and come in two basic compositions, tissue-core, and film. Tissues have a center core or carrier of either porous tissue (ColorMount, TM-2, Trimount, "Super" Unimount, Promount...) or nonporous glassine type sheet (MT-5, TM-1, Postermount...), with adhesive applied to either side of the carrier for mounting.

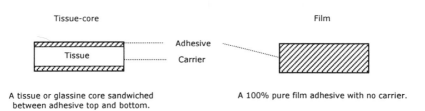

A tissue or glassine core sandwiched between adhesive top and bottom.

A 100% pure film adhesive with no carrier.

Since both sides are identical, there is no top or bottom. They are clean, dry, nontacky, relatively opaque white in color, and are also extremely time effective for production use. Tissues adapt extremely well to oversized mountings, float mounting or multiple bite procedures.

Pure film adhesives (Fusion 4000, TM-3, Flobond, Acid-Free Mounting Film, Versamount...) are 100% adhesive with no carrier or tissue in the center. This makes them translucent when unmounted and clear when mounted. Some films may also be pieced or overlapped because of the lack of central carrier paper, which allows for greater use of scraps.

DEGREE OF POROSITY (BREATHABLE OR NONBREATHABLE)

Porosity is the level of which an adhesive is permeable by moisture or air. This is an extremely important designation when selecting a tissue for compatibility with all selected mounting materials. If a nonporous/nonbreathable material, such as a photograph or heavily lacquered print is to be mounted, the adhesive must remain breathable to allow for air or steam to be forced out and/or through the mounting layers. If this is not allowed the project will suffocate.

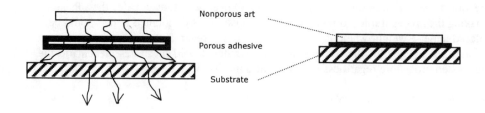

By using a nonbreathable adhesive with a nonbreathable photo, there is much greater potential for air to be trapped between the two nonporous items creating bubbles in the completed mounting.

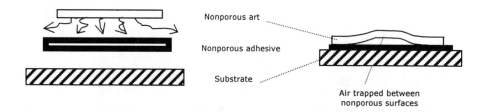

If a breathable item is to be mounted to a breathable substrate, essentially any tissue may be used to mount it simply because air will always be able to be compressed out through and around the porous art and substrate.

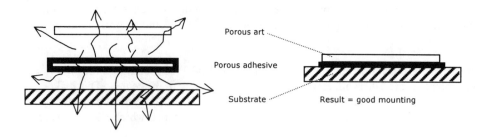

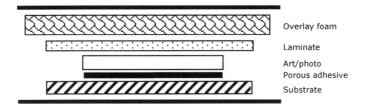

Overlay foam
Laminate
Art/photo
Porous adhesive
Substrate

In any mounting or laminating package, only one layer within the sandwich between release materials on top and bottom may be nonporous. If two, such as photo and adhesive or photo and nonperforated laminate, exist air is likely to remain trapped between the layers in the completed mounting. In other words, the project will suffocate. Only one layer in any mounting or laminating package may be nonporous, all others must be allowed to breathe. The technique of perforating a nonporous laminate temporarily allows it the porosity necessary to be used over a nonbreathable photo. This prevents two-layer suffocation.

ACIDITY LEVEL OR pH
Since most dry mount adhesives are inert, meaning they contain no harmful acids capable of damage, it is actually the carrier sheet that needs to be checked for pH levels. Many manufacturers have developed tissues using acid-free or archival carrier papers in conjunction with dry mounting adhesives and have named them accordingly. These tissues mount at lower temperatures, are breathable, removable and neutral pH. They are considered more delicate, but **do not meet conservation standards**.

Using heat-activated adhesives can never be considered archival because the very act of dry mounting art to a substrate breaks all conservation guidelines. Adhesives travel toward the heat as they are mounted, and dry mounting encourages a certain percentage of nonreversible adhesive to penetrate the back of the artwork, even with a removable tissue.

Dry mount adhesives are available from a number of manufacturers and distributors as both name brands and private label brands. Adhesive prices do reflect the thickness of tissues and the amount of applied adhesive per side, so price comparisons need to reflect an understanding of variations in the physical nature of the products.

PREADHESIVED BOARDS
Adhesives already applied to mount and foam boards are found as both pressure-sensitives and heat-activated types and are available from a number of manufacturers. The heat-activated boards come with activation temperatures of between low 150°F to average 190°F. The substrate structure, rigidity, and ease of sizing all remain consistent as with any mount board. Designed specifically as a convenience and time saver, the adhesive/board pricing generally reflects the two materials purchased separately.

SUBSTRATES

A substrate is a backing (board, canvas, or other media) used to affix or mount artwork to. Selecting the right board as a substrate is as important as selecting the correct adhesive and mounting procedure for any specific piece of artwork. It is terribly important to be consistent with material selection throughout the entire job, meaning follow through with the chosen concept. If acid-free mats are chosen, then acid-free mounting substrates and neutral pH adhesives should be used. An acid-free or inert adhesive on an acidic substrate still equates to potential acid burn. Just as a non-neutral pH tissue on acid-free foam is futile.

Almost any rigid surface may be used as a mounting substrate, including glass, foam, aluminum, Masonite, or paper ply boards. Europeans continue to use a great deal of MDF (medium density fiber) board and hardboard for framing, which is much less common in the US, though chipboard and corrugated boards may still be found. In order to best eliminate warping, the weight, thickness, and rigidity of the selected board should be adequate to accommodate the art being mounted. To prevent as much bowing as possible, select thicker harder boards for larger artwork, as this helps compensate for greater surface stresses. Though a board product, MDF is rather soft and prone to bowing under surface stresses.

Other considerations when selecting a substrate include the higher acidic levels when using regular mount board, chipboard, greyboard, MDF, hardboard, Masonite or plywood; and the degree of orange peel which might appear with glossy photos and some poster art which reflects the board surface beneath.

TRADITIONAL BASICS

Whether used for mounting or as a filler board, rigidity is the main purpose for backing boards or substrates. When dry mounting, boards undergo the highest degree of prolonged pressure, making them more susceptible to crushing. Foam boards will not physically melt until they reach internal temperatures of 230°F, but will compress around the outer edges from the pressure exerted during mounting. If uncrushed square edges are required, it is necessary to mount on a slightly oversized board then trim down to the desired dimensions after mounting. Compressed outer edges of foam in no way effect the actual mounting and just are a natural end product of using heat mounting vacuums.

The following thicknesses are suggestions for standard mountings:

Size	Board
Up to 8x10"	4-ply mat board, x board, or equivalent
8x10" to 16x20"	1/8" foam board, 3x board, MDF, or equivalent
Over 16x20"	3/16" foam board, MDF, or equivalent

Additional rigidity of the substrate such as heavy-duty foam boards or hardboards are suggested for projects over 32x40", especially when being mounted for nonframing or when they will be subjected to a high humidities.

MAT BOARDS
Warping of large pieces is a common problem with 4-ply mat board and flawboards. These should probably not be used as a mounting surface if the print is any larger than 16x20". Using too thin a substrate for too large a mounting will encourage warping of the mount board. The board may be countermounted with something similar on the back, but it is better to select a heavier board for larger pieces.

Countermounting is the process of mounting a piece of paper similar to the art on the reverse side. This reverse mounting is designed to create equal tension on both surfaces of the substrate, neutralizing the tendency to bow. See Wet Mounting, page 40.

FOAM BOARDS
The lightness, rigidity, and ease of cutting, are what foam board has become known for. Foam boards are available as regular clay-coated, colored surface paper, acid-free, and 100% cotton rag. Regular boards are available with both white and black core foam, and a variety of both white and black surface papers in an assortment of sizes in 1/8", 3/16" and 1/2" thicknesses. This extensive selection allows for choosing the correct board to fit each framers need. Foam boards are probably used more than any other board for mounting in the United States as MDF, pulpboard, greyboard, or hardboard is in Europe.

Acid-free foam boards are also frequently used as filler or backing board behind hinged museum 4-ply boards in conservation applications. They are noted as the current foam board of choice. Using acid-free foam creates more of a consistency of selected materials in a framing package if all the mats and boards are acid-free or acid-buffered. For example, the 100% cotton rag foam by Bienfang meets surface paper conservation quality suggestions for backing for either hinging or mounting.

Black surfaced foam boards work well for color control when color tinting or accenting the edges of unframed bevels. It is a great choice for posters with exposed bevel edges to give a more finished look to the mounting when a frame is not used. The black of the surface paper may transfer to light colored walls as it ages leaving a black echo when removed. Laminating the back side of the unframed project will act as a barrier to protect painted walls from the echo, as well as the benefit of countermounting in the process.

Foams with colored surface papers are good for controlling ghosting from lettering on the opposite side. By selecting the mounting surface to match the dominant verso color, ghosting is drastically reduced or eliminated.

HEAVY DUTY FOAM BOARDS
The biggest difference between the heavy-duty foams appears to be the toughness of the inner core vs. the toughness of the outer coating. Price and ease of cutting varies also.

MIGHTYCORE: Hunt Corporation has a foam product with a heavier and stronger polystyrene core, and smooth moisture-resistant polyester-coated surface paper which resists denting, bending and scratching. Claiming strength and smoothness, it could be a good choice for photos, but be careful of the nonporous nature of the surface. Available in 1/4" and 1/2" thicknesses, in standard board sizes and six colors, CFC free.

NUCOR: Savage Universal offers Nucor, a dense polystyrene core board laminated on both sides with polycoated facing papers. The inner core and face papers make a moisture- and warp-resistant product well suited to dealing with humidity. Available as plain, heat-activated, and nonrepositionable pressure-sensitive boards, 3/16" thick. Consider porosity when selecting substrates.

GATORFOAM: Much tougher than a standard foam board is Gatorfoam from International Paper. It is rigid polystyrene foam with wood fiber veneer applied to either side making it stronger and more durable than other foam products. It is extremely smooth, rigid and resistant to warping, making it perfect for oversized mountings. One drawback is its difficulty in cutting. This may be overcome when using certain adaptable heavy-duty wall cutters designed for cutting hardboard such as Masonite and Gatorfoam. It is available in white, natural and black in thicknesses from 3/16" to 1-1/2" in 4'x8' sheets.

OTHER BOARDS
In the commercial market, very large oversized posters can be a challenge to mount and selected substrates may include Masonite, Abitibi board, Beaverboard, Gatorfoam, plywood, Plexiglas, or Sintra. These boards may eliminate warping and bowing of substrate, but other issues need consideration. Though they mount in a heat press, are best for use with rollers. When vacuum mounting a two-step method should be initiated to compensate for the lack of porosity. As with heavy woods, plastics might need to be professionally sized, and will demand a higher mounting price. Moisture is a problem as well as porosity. Make sure the woods are very dry prior to mounting.

CORRUGATED BACKING BOARDS
Boards used to fill the space between the substrate and the dust cover are backing or filler boards. Most commonly used are foam and corrugated boards. Note that mounting onto corrugated card or poly board can create a series of ridges under the artwork, not the best visual choice and may be acidic. Including any acidic materials in the sealed frame will allow for acids to migrate to the artwork, regardless of placement in the package.

Corrugated materials range from basic inexpensive brown cardboard to heat resistant plastics of a water-repellent nature. There are blue-gray corrugated acid and lignin-free, buffered boards well suited as spacers or for conservation backing. Inert corrugated plastics or polyflute (polypropylene) boards used in conservation are not appropriate for mounting as they are nonporous and have ridges that would show.

PHOTO BOARDS

Archival photo boards are nonbuffered 100% cotton fiber museum boards, acid and lignin-free, neutral pH, manufactured for art and photos requiring a low alkaline level. Some are a 4-ply created by bonding 2-ply to 2-ply with an adhesive polymer between, making them rigid enough to meet museum requirements. Available generally in neutral colors of cream, antique white and white, they meet Library of Congress standards and the requirements set by photo conservators as mounting substrates for photos. All mat board manufacturers handle their own version of nonbuffered photo boards. Check specifiers for details. Often these boards need to be special ordered through distributors.

FABRICS

At times, linen and cotton fabrics may be selected as a reinforcement for aging and antique art, making them a substrate. Artwork may be mounted with wet glues or dry mount adhesive films when used for antique art, such as old movie posters. It is recommended to leave the mounted paper-to-linen unit unstretched so the two materials are allowed to expand with relative humidity and temperatures. The two different fiber types will react differently to outside influences and when stretched onto bars the restrictions of movement could encourage delamination (bond failure).

Raw and heat-activated canvas fabrics are used as the substrate for canvas transferring, also known as canvas bonding. See Chapter 12, pages 123-132 for additional information on canvas.

Synthetic fabrics may be selected as a substrate or substrate cover for decorative art. The unstable nature of these fabrics often makes them light fugitive, meaning they may fade. The acidity level will vary with the dyes and specific fibers selected. Fabrics of unknown origin are not the best choice when framing valuable artwork, as they are probably nowhere near conservation level.

Neutral pH fabrics of cotton, linen and canvas are now available for wrapping, backing and shadow boxes when used for conservation/preservation framing. Many fabric surface mat boards have also been released for conservation use, claiming neutral or inert fabrics have been used in their construction.

> ***TIP:** PEC 12 (PHOTOGRAPHIC EMULSION CLEANER)*
> When water does not work, PEC 12 is a solvent and cleaner available at photo supply stores that can remove bits of adhesive, smudges, and sometimes fingerprints from the photo surface.
>
> DO NOT USE on digital photos as the inks will dissolve and smudge.

RELEASE MATERIALS

In mounting, release materials are designed to keep adhesives from contaminating mounting surfaces, and in turn to keep adhesive residues from coming into contact with artwork. They protect both the heating platen in the top of the press and the pad or diaphragm in the bottom of the press. There are many types of silicone-coated papers and films, all with varying degrees of nonstick capability.

Release materials may be coated or uncoated having inherent release properties such as Teflon or polyethylene, which prevent adhesive absorption, making them easily removable from adhesive surfaces (peelable). Some are used as removable liners to protect pressure-sensitive or heat-activated adhesives until ready to use. Others are applied to the back of laminating films to allow for rolling and storage. They come in a variety of sizes and are available both one-sided and two-sided. Translucent Mylar or polyester films are also available for a smooth release sheet that can be seen through (see Clear Release Films, this page).

Indentations, or wrinkles, are created in bottom release papers and films when mounting substrates are properly pushed down into a sponge pad of a mechanical press, or the diaphragm sucks up around the substrate in a vacuum press. They are the natural result of pressure during the mounting process. Check routinely, and retire overly wrinkled release materials for new ones.

CLEAR RELEASE FILMS

Clear release paper is not a paper at all but a coated sheet of plastic film. It was originally designed to be used in conjunction with glass topped vacuum presses so mountings could be better monitored. Rolls (as the paper) and sheets are available up to 50x102". They are smooth and sometimes recommended for use with gloss laminates to reduce gloss damage during mounting.

DOUBLE-SIDED RELEASE PAPER

Two-sided release paper is a slick, lightweight white paper with a silicone coating on either side. The base paper is lightweight making it tougher to handle when dealing with oversized mountings. Its slick, limp and more prone to folding over when working with a 40x60" format. The lower price and flexibility of using either side of double-sided paper makes it a big seller.

SINGLE-SIDED RELEASE PAPER

One-sided release paper is made of a heavier base paper, the silicone coating has a pale blue appearance, and the opposite side is a matte, uncoated white paper. It is more rigid and nonslippery, making it perfect for wrapping around the sponge pad of a mechanical press, or hardboard carrier of a hardbed press. It is also the only release material that can be bonded to plain board for making in-store release boards.

RELEASE BOARDS

These are one-sided release papers bonded to a thin substrate board. Release boards are commercially available up to 40x60" in size, or you can make them yourself using a baseboard and a permanent, porous tissue adhesive, one that bonds under heat in the press. They are good for an average of 50 hours work time before needing to be replaced.

Release boards are optional only being required for use during multiple bites in a mechanical press. The board will help dissipate the pressure at the outer edge of the press plates to better prevent dents when mounting onto a foam substrate. They are only a convenience for use in a vacuum press, never required.

Though release boards do not easily indent, wrinkle or fold over, do not use them both top and bottom in either a mechanical or a vacuum system. In a mechanical press, the project needs to nest down into the sponge pad to ensure a constant, even pressure against the platen. In a vacuum press, the bladder or diaphragm needs to be allowed to contour up around the substrate in order to adjust its pressure against the platen. If a board is used both top and bottom, excess pressure can occur at the outer edges as the sponge or diaphragm attempts to conform to the shape of the desired inner project. The release board can then possibly create uneven mounting pressure in the project center.

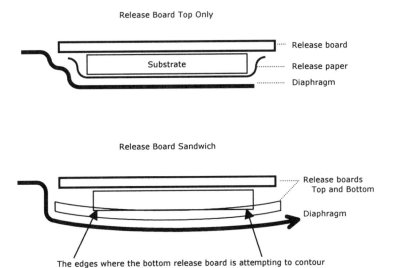

The edges where the bottom release board is attempting to contour around the substrate has a greater degree of pressure than the inner center of the mounting itself...air bubble potential.

SOLVENTS

Substances that breakdown or dissolve another substance are known as solvents. The most common solvent is water. Evaporating solvents are used in paints, varnishes, and plastic coatings to keep them liquid. Once those solvents evaporate, the liquid solution becomes a tough solid. Bond failure can occur when a thermoplastic adhesive is rewarmed in a press, such as long time bonded pressure-sensitive tapes and spray adhesives. The heat reactivates (resoftens) the adhesive returning it to a liquid allowing for easy separation, though adhesive residue remains soaked into both surfaces.

Even permanent dry mount tissues may be removed by using specially formulated solvents. They are strong, hazardous chemicals that evaporate quickly when exposed to open air and readily dissolve most dry mount, spray, and pressure-sensitive adhesives. These include commercial adhesive release, Bestine thinner, mineral spirits, toluene, and acetone. Fumes are present when using solvents, use them only in well-ventilated areas.

Solvents could be chemicals that might damage the artwork attempting to be salvaged. Testing of the inks and dyes is mandatory before submerging or utilizing any solvent. They may also be used to clean photograph surfaces of fingerprint oils and to clean press platens of alien adhesives. The key is to use common sense.

SOLVENTS TO REMOVE PERMANENT ADHESIVES
MATERIALS
UnSeal, Bestine thinner, Toluene etc.
Photo developing tray to fit substrate
   Or as a temporary tray for larger pieces
Stretcher bars and plastic garbage bag as liner

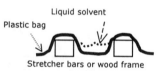

1. Check printed image for ink stability with Q-tips and solvent,
   by soaking tip in solution and lightly touching separate inks on print.
   If photo, all colors are permanent, simply submerge into solvent.
2. Pour liquid into tray.
3. Submerge safe art or photo and soak about 5 minutes.
4. Remove from solvent, it will air dry almost immediately.
5. Separate photo from adhesive/substrate, using a sharp blade tip, saturated adhesive removes easily, or it could be floating on the surface already.

> **TIP**: *SOLVENT SENSITIVE INKS*
> With solvent sensitive inks, droppers of solvent might need to be flooded a little at a time between print and adhesive working from a diagonal corner. This keeps the solvent from contacting the surface inks. Time required for adhesive removal with a dropper might be better spent by replacing the poster instead.

REMOVING ADHESIVES
Any time an item is glued down by any other than starch pastes, the conservation concept of an archival presentation has been violated, you may never truly bring that item back to its original state. **Archival treatment of fine art refers to the relative permanence of materials used, their ability to remain stable over time, and not altering the original state of the art.** Mounting with many wet, spray, pressure-sensitive, or dry adhesives are nonconservation adhesives. **Removable does not mean reversible!**

TAPES
Some tapes are water-soluble, some require chemical solvents, and some dry up and fall off in time. Check with individual manufacturers for removal techniques. Two additional methods include applying heat to the tape with a tacking iron to reliquify the adhesive, or lighter fluid brushed on in a small amount, will often dissolve adhesives.

Rubber based adhesives used in masking and cellophane tapes, although considered natural, are prone to yellowing. Acid-free products claim to remain nonstaining and nondamaging over the test of time, but it may take years to accurately verify their claim.

WET GLUES AND SPRAYS
As sprays age, the bonding agents holding them to the substrate dry out and may begin to peel from the substrate. If a partially peeled poster has been returned for remounting, placing the partially mounted image into a 180°F press for a few minutes will often activate the adhesive making removal a simple peel from the board.

As with dry mount tissues, many adhesives for hand mounting and cold vacuum frames fall into similar removable and acid-free categories. Check manufacturer guidelines for mounting suggestions and pH levels, then let your conscience be your guide.

> ➤ *TIP: ALTERNATE HEATING or REMOVAL BY HAIR DRYER*
> Some water-soluble wet glues, as well as removable HA adhesives may sometimes be softened for easier removed by heating the surface of the art with a hair dryer or shrink-wrap gun to reactivate its thermoplastic adhesive.

DRY MOUNT TISSUES AND FILMS
The procedure of removing an item once mounted to a substrate leaves traces of adhesive residue behind. Though most adhesives are inert (neutral pH), even a tiny amount of adhesive which may have soaked into the paper or fabric has altered that artwork from its original state.

Once removed from the substrate, enough adhesive may have saturated into porous art for the residue alone to bond it again to a new substrate with no additional adhesive tissue. Solvents may remove some of this adhesive, but color tests must be done to ensure the permanency of the inks and pigments of a print.

HEAT-ACTIVATED TISSUES FROM PRINTS

Removable tissues and films are reactivated by putting them back into the heat of the press, *slightly* warmer for a *little* longer than the recommended mounting procedure. When adequate time (5 minutes longer) and temperature (10°F hotter) have been met to reliquify the adhesive, remove the project from the press and separate the artwork from the adhesive leaving the adhesive tissue still mounted to the substrate and the artwork free of the tissue. If archival tissues are placed in the press **too hot, too long** the adhesive excessively absorbs into the substrate and artwork making it extremely difficult to separate the remaining tissue from the back of the art.

Since adhesives saturate toward the heat source, a possible precaution might be to reheat the project for removal face down. The adhesive will then be more likely to pull into the disposable substrate than the print being salvaged. There will be adhesive residue remaining on the back of the artwork, which can be removed with liquid solvents. To remove the maximum amount of adhesive without the use of solvents, continue to remount and remove the art until it will no longer hold to a new clean substrate.

PURE FILM ADHESIVES

The above removal procedure is the same for 100% pure film adhesives with no tissue core. Simply reheat the mounting a little hotter a little longer then separate immediately upon removal from the press. Adhesives that bond as they cool set up quickly once removed from the heat source, so separating them immediately is imperative.

PERMANENT TISSUES FROM PHOTOS

Slick paper stock and RC photos resist adhesive absorption so removal is quite easy, but textured and rag paper stocks may cling aggressively to removable tissue. Permanent tissues may be removed even though they are called permanent. Solvents used for this may not be acid-free, and are quite toxic. Saturate a Q-tip with the solvent and touch it to the various colors on a print, as you would test needlework for color permanence.

Small images may be submerged into the solvent solution for about 5 minutes to dissolve the adhesive, then separate the image from the tissue and substrate unit. The solvent evaporates rather quickly limiting the mess factor but you need to work fast. It may also be saved for future use in another bath by pouring it back into the original container. Be sure to mark the contaminated can not to be used for cleaning the platen or photos.

3

Setup and Troubleshooting

CLEAN AREA...CLEAN PROCESS
Though the location you choose for mounting is not directly a basic element to ensure successful mounting, it effects efficiency. Careful thought and planning should be paid to the work area, workspace or mounting station where all of the vital mounting elements are to be controlled.

Organization often lends itself to a clean workspace. You must organize the mounting area to accommodate your specific requirements. If dealing with sprays, a well ventilated and masked spray booth is required, or if working with a mount press space requirements include accessibility. Cleanliness, or lack of, will often transfer directly to the art. Dust and debris circulating in the air may become trapped under the mounted art or fabric. Try not to set up your mounting station near any other dust creating framing process. A miter saw will throw metal and wood chips into the air, a heavily used wall cutter may create paper dust and particles, and even glass particles could be a problem. Tiny pinhole indentations usually indicate dirt on the platen or release paper surface, while bumps are dirt trapped between the print, adhesive, and substrate.

Proper tools and equipment should be ergonomically placed within reach for easiest execution. Lost time looking for misplaced tools or removing unwanted dust from beneath a tissue adhesive translates into burned up profits.

WORKSTATION
LIGHTING
Dust and fibers will never be detected if they cannot be seen. Working in a darkened environment is not only a headache, but also can be depressing, creating frustration and eyestrain when the lighting is inadequate. Shadows produced by bad lighting must be eliminated.

A noncorrective fluorescent tube light fixture may not give you the color reference of natural light, but it can efficiently illuminate the work area. Release materials need light to be examined for adhesive residue, wrinkles, folds, and potential problem troubleshooting prior to each mounting. Particles cannot be removed from between the print and substrate if they cannot be seen.

STORAGE
Mounting board storage is best when kept clean, dry, and flat, though most storage facilities stand boards on end. By standing them up, they are encouraged to warp during storage, which only adds to the warping frustration that can be created when mounting large pieces on lightweight boards.

If boards are stored in a basement, warehouse or other room subject to extremes of temperature and humidity, the issue of predrying these boards prior to mounting may carry over into vacuum mounting systems where predrying is generally not required. Most vacuum frames and presses are set up with an optional lower shelf, which makes an ideal storage space for boxes of foam, and mount board. Sheets of ½" foam board make great rigid space filler to help prevent warping of vertically stored boards.

Dry mount adhesives should be stored in a clean, dry and accessible locale. They should be well labeled to avoid mix-up and be stored away from release papers for the same reason. A convenient storage or dispensing rack to the side of your mounting press is ideal. This way dry adhesive can be pulled to the desired length, cut, positioned, tacked and mounted with no confusion or clutter.

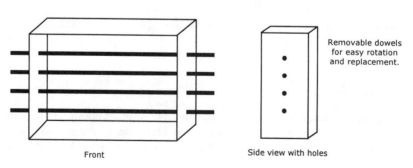

Adhesive rack (Not to scale or proportion)

Front   Side view with holes

Removable dowels for easy rotation and replacement.

PLACEMENT

The actual location of the equipment will vary as to whether you have a vacuum, mechanical or hand application system as well as workspace limitations and ventilation requirements. If setting up a mechanical press, the ideal situation is to recess it into the worktable so the sponge pad is even with the table surface. This will allow easy level insertion into and out of the press.

The mounting table should have a cover sheet of clear plate glass. Glass is excellent to use as a cooling weight, as it may be seen through, and used as a cutting surface. This glass should be large enough to accommodate most of the anticipated mountings, but still small enough to handle, usually no larger than 32x40".

Requirements of an efficient workspace include:
- Clean area
- Adequate lighting
- Organized storage of adhesives and substrate
- Placement and height of equipment
- Cooling table and weight

Also, remember water for clean up if using wet glues and electricity for tacking irons. Drawers for small tools and miscellaneous materials are also an asset in the preparation table.

Maintain adequate space around a mechanical press to allow the benefits of oversized mounting in bites. When designing a new workspace try to imagine all of the extremes, such as multiple production mountings, oversized mountings, laminating, and creative applications. Keep materials handy yet out of the way and be efficiently as productive as possible.

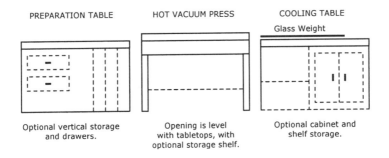

PREPARATION TABLE
Optional vertical storage and drawers.

HOT VACUUM PRESS
Opening is level with tabletops, with optional storage shelf.

COOLING TABLE
Glass Weight
Optional cabinet and shelf storage.

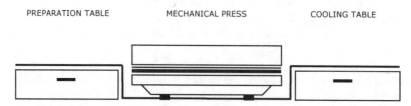

Top of the mechanical press sponge pad should be set level and aligned with the tabletops on either side for easy entry into or exit from the press.

MECHANICAL PRESS MAINTENANCE
DAILY ROUTINE

Develop a daily routine of wiping down the inside of your mounting equipment to cut down on particles looking for a place to relocate. In addition, keep presses closed when not in use to avoid dust and particle buildup. Just as with your mat cutter, daily cleaning and nightly covering will allow for better control over your work.

- Keep the press closed, but not locked, when not in use and cover nightly to avoid dust and particle accumulation.

- Check for adhesive residue and scratches.

- Avoid tiny dust pits and indentations in completed mountings by regularly wiping release materials with a clean, soft, lint-free rag to remove bits of unwanted adhesive and dust particles.

- Good lighting is important to be able to see alien particles; they cannot be removed if they are never seen.

PLATEN CLEANING

Adhesive buildup can be removed by heating the press to 200°F, and turning it off with a piece of clean Kraft paper clamped closed inside overnight. As the press cools most of the adhesive will transfer to the paper. Remove the paper and discard in the morning.

If additional cleaning is necessary, apply cream platen cleaner to a cold press and scrub with a nonabrasive pad to remove residue. Available through distributors. Open windows for ventilation when working, and never use blades or sandpaper that could scratch the platen surface.

UNHINGING A MECHANICAL PRESS

If the platen of your mechanical press is in need of massive attention due to lengthy neglect or long term buildup, the hex bolts and nuts located on the lower arm of the press may be removed allowing a full opening of the press.

Always remove the lowest of the three bolts and be careful to inspect the placement and exact order in which the washers, bolt, and nuts need to be replaced. Once the lower bolt has been removed (both 9/16" and 5/8" are required for the 210M in the photo) and the top is laid open like a book, platen accessibility for application of solvents and creams is quite easy.

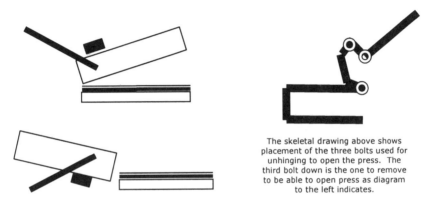

The skeletal drawing above shows placement of the three bolts used for unhinging to open the press. The third bolt down is the one to remove to be able to open press as diagram to the left indicates.

RELEASE WRAPS

Wrapping single-sided release paper around the sponge pad of a mechanical press will protect it if a project were to be accidentally mounted without a release envelope. This could allow edges of adhesive to be physically mounted to the felt layer on top of the pad.

Cut a piece of release paper to the width of the pad, fold the excess paper around the ends of the pad and tuck them <u>between</u> (not under) the pad and the Masonite. The limp heavy nature of the pad in conjunction with the toothed back of the single-sided release paper holds well without much slipping. A wrapped pad is not meant to be a substitution for release materials, but rather additional insurance to protect the pad from ambitious new mounters who may forget the critical release paper bottom sheet.

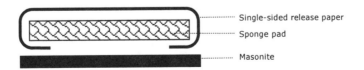

Single-sided release paper
Sponge pad
Masonite

VACUUM PRESS MAINTENANCE
ADJUSTMENTS
Vacuum presses need little adjustment and maintenance once they are set up and running. Individual manufacturer manuals will reinforce all step-by-step maintenance procedures for their presses. The location for the press should be level and the press should be checked for level once installed.

If everything appears level but there is not adequate vacuum, check the following:
> the lid may be out of alignment to the press,
> there could be an air leak in the outer foam cushion, or
> release materials may simply be invading the outer edge seal.

If the lid is out of alignment, read the manual to see if the press has hinge adjustments in the back. Loosen the hinges, draw the vacuum, retighten the hinges and it may be corrected.

The vacuum portals of a press may be located at one rear corner, both corners, or center of the press depending on the manufacturer. They draw the air from within the press, so must be kept clean, and in some cases covered with the felt liner in the bottom of the press, to work properly.

Fittings running to and from the press rarely need readjusting but initial tightening is important. Rubber bladders or diaphragms are the soft moldable base in the bottom of the unit that conforms around the mounting substrates to press them against the heated platen or glass. They may be textured or smooth and vary between 1-1/2" to 3" in relaxed drop when not in operation. Some press motors require routine oiling, or occasional filter cleaning, check owner's manual for maintenance details.

DAILY ROUTINE
In moist or high humidity areas hoses may become blackened with residue from excessive moisture from mounting. Even with the following daily routine, hoses may remain black:

- Run an empty press first thing in the morning through one full heated cycle to blow out the hoses and draw out any accumulated moisture from within the unit.

- The final press run of the day should also be empty, but with the press lid open to draw room air through the vacuum hoses.

CLEANING THE VACUUM PLATEN
If examination of the platen through wiping with a lint-free rag or by running your hand smoothly across the cold surface indicates particles or adhesive, it is imperative to remove them. Using an adhesive solvent such as UnSeal, un-du, Bestine thinner, or toluene easily removes adhesive, paper and foam residue. Make certain to open windows for adequate ventilation. Never use any abrasive materials on the platen, sandpaper, steel wool or sharp objects could permanently scratch the platen surface allowing these permanent scratches to be transferred to later mounting projects.

A commercial cream platen cleaner comes with a nonabrasive pad specifically designed for removal of stubborn adhesive residue. If all attempts at cleaning fail, perhaps the platen should be checked for scratches, it may require refinishing or replacement. Call manufacturer directly for details and advice.

Silicone residue from release materials often results in a visual discoloration on the surface of used vacuum platens. This is generally detectable only when viewed at an angle and poses no danger to successful mountings since there is no texture to transfer indentations to the project.

CLEANING VACUUM PRESS GLASS
Commercial solvents will dissolve adhesives on glass just as on platens. Since generally the coating for heat distribution is on the glass side, away from the inside of the press, routine glass cleaners, lighter fluid and glass (window) scrapers may be used to assist in cleaning the glass. Check with manufacturer for maintenance suggestions.

SUGGESTED MOUNTING METHODS
See appendix, pages 136-137.

> ➤ ***TIP: POLLUTANTS***
> It is becoming increasingly clear that all forms of pollution (including: fumes from fresh paint, new carpeting, household cleaners, air conditioning, pesticides, cooking, smoke, dust...) are the greatest enemies of artwork. Many products outgas contributing to the damage. Selecting materials that are neutral pH and/or pollutant protective will help retard inevitable damage.

4

Elements of Mounting…TTPM

Time, temperature, pressure and moisture (TTPM) are the four basic mounting elements from which all successful mountings will develop; and in opposition, reasons for any unsuccessful mountings may be tracked. All mountings may be individually analyzed by going back to these same four basic elements. Time, temperature, pressure and moisture will vary depending upon the mounting process, equipment and adhesive selected.

TIME

The basics of wet, spray and pressure-sensitive mounting all follow TTPM procedures, but consider the element of time as *tack time* (also called *open time*) rather than literal time on the clock. This is the workable time allowed by an adhesive to position the print being mounted as the solvent evaporates in preparation for final set-up. Open times vary depending on the product so be familiar with its proper usage.

Application time is taken into account with wet and spray adhesives in that a good uniform coat of adhesive must be applied in order to guarantee good adhesion. Any areas that have begun to dry prior to mounting the art will not create a lasting bond. Although a basic bond will be created within the first hour, more permanent bonding will take 4-24 hours and should be done in an undisturbed area under pressure.

Open time is the time required for a spray adhesive to allow solvents to dissipate from the adhesive so that it may become tacky. Not allowing adequate open time could effect the long term bonding nature of a potentially successful spray mounting.

The time it takes to dry mount a project will also vary depending upon the adhesive, substrate, mounting size, selected temperature and item being mounted. *Dwell time* is the time remaining in the press to adequately heat all inert materials, activate the adhesive, and create the bond.

If a project is removed too soon for proper bonding it may be repositioned in the press for a second time. The second time in the press it must remain for the initial mount time PLUS the additional:

    3   initial minutes first time in the press
 + 1/2   additional minute desired
 3-1/2   total minutes second time in the press

The most important thing is to match the proper time/temperature ratios. It is better to work at lower temperatures with a longer dwell time than to increase temperatures in an attempt to speed up the production process. It's also better to slightly over-estimate desired press time than to have to place it back into the press to achieve proper bond.

MECHANICAL PRESS
The overall size and thickness of the selected substrate will determine adequate press time. Anywhere from 1 minute for an 8x10" to 3 minutes for a 20x24" is average time. Predrying is also required adding another 10-15 seconds per item.

VACUUM PRESS
Total cycle time for an average mounting is 4 minutes for a 32x40" board with 5-6 minutes for 40x60". *Draw time* is the length of time it takes for all of the air to be sucked out from between the heating platen and lower diaphragm within a sealed vacuum press. The draw times will vary with the depth of the relaxed diaphragm, strength of the pump, and press size. Some presses will draw as quickly as 15 seconds and mount in 30 seconds while others will draw in 1 minute and mount in 4. Check individual manufacturers' manuals for expected draw and overall mounting times. Predrying is not required in a vacuum unit.

TEMPERATURE
Predictability is the aim when mounting. When repeatedly mounting similar items the results should be exactly the same, and practice makes perfect.

WET AND SPRAY MOUNTING
Recommended storage and room temperatures are suggested on all bottles and cans. Frigid temperatures restrict shipping of many wet glues due to freezing, and generally temperatures below 60°F are undesirable for mounting applications. Temperature extremes and cold tend to stiffen and solidify flow and encourage glues to clump.

DRY MOUNTING
Temperature is the only element that lends itself more to dry mounting than the rest. And all adhesives will have manufacturer's suggested temperatures for use in order to achieve best results. There is not a standard, ideal temperature to be used in every heat-mounted situation, for dry mount adhesives range from 150°F to 225°F.

Individual tissues and pure adhesive film temperatures vary depending upon the item being mounted and its substrate thickness. Establishing a basic average temperature for most daily mounting will cut down on excessive press fluctuation, which saves time and money. Generally, 185-190°F is an average temperature for most dry mounting.

TIME/TEMPERATURE RATIOS
Though it may take 4 minutes to mount at 180°F, it is not encouraged to play with the temperature to speed up the process. At 200°F it might mount quicker but not as safely as at the lower 180°F. It is better to be slow and safe than fast and sorry.

PRESSURE
The successful process of mounting consists of applying adhesive to art and substrate, then allowing it to harden under pressure. Pressure is the force that squeezes the air from between the substrate, adhesive, and artwork being mounted and holds it while the bond is created. Whether wet, spray, or dry mounting, the practice of weighting needs to be applied in all mounting cases.

WET AND SPRAY MOUNTING
Use of pressure during the setup process of wet or spray mounting encourages a good bond and helps reflatten the moistened substrate as it dries. This pressure may be as simple as a piece of 1/4" plate glass or as complex as a cold vacuum frame, but is a somewhat variable portion of successful hand mounting.

There is no designated appropriate poundage established for wet glue pressure to be effective, it remains simply *good technique*, to maintain maximum contact while the glue dries and sets up.

PRESSURE-SENSITIVE
Both preadhesived pressure-sensitive boards and two-sided adhesive films such as PMA require pressure to properly bond. The major difference being these adhesives do not need to remain under pressure as they dry. They actually require the pressure of a squeegee, rubber roller, or roller machine to activate the adhesive and create the initial bond. Although a basic bond is established almost immediately, the most long lasting bond occurs after 24 hours.

DRY MOUNTING PRESSURE VARIATIONS
Both hot and cold vacuum presses are self-adjusting in relation to the substrate or mount board being used. The rubber diaphragm or bladder, which forms the bottom of the unit, naturally conforms to the thickness of each individual substrate during the draw of the vacuum, adjusting for pressure automatically.

They never need to be manually adjusted. Dry mounting pressure required in a mechanical press is very specific. Inadequate pressure might allow air bubbles to remain within the center of a mounting, while too much pressure could create unsightly indentations in a foam board substrate during a multiple bite project.

Average pressure poundage in a vacuum press is somewhere between 11-15 pounds per square inch (psi) while pressure in a properly adjusted standard mechanical (softbed press) is 2-4psi. Hardbed presses, used most often outside the United States, have a much higher range of psi pressure potential, though averages around 5psi for average wheel rotations.

Advantages to using a vacuum press include the flexibility of the diaphragm (or rubber base) to conform to the variable substrates being used. If mounting a 4-ply mount board prior to a 1/2" foam board and then 1/8" foam board, no adjustments need to be made with a vacuum press. If these variations were needed when using a mechanical press the pressure arm would need to be adjusted each time to create proper tension for adhesion. A mechanical press must be manually set to apply the appropriate pressure for the thickness of substrate.

MOISTURE
Make sure the **time** is correct for the thickness, size and tack for the materials being mounted, the **temperature** is set to accommodate the type of adhesive being used, and that appropriate **pressure** is applied for the above combinations. As with the others, **moisture** is an element that also varies with the process selected for mounting. All wet and spray adhesives contain water, so moisture is good. Any moisture that is present during dry mounting can become a condensed liquid or steam, and this type of moisture is very bad.

Moisture is easy to understand and control, and by doing so will ensure repeatedly well executed, smoothly mounted, long lasting projects. Knowing when to think dry and when to bend that rule with dampness is the key.

WET AND SPRAY MOISTURE
When working with wet adhesives, moisture is an integral part of the adhesive. Selected mounting materials must be allowed their natural expansion of when moisture is applied to dry paper or substrate.

A precaution for dealing with paper expansion and wet glues is moistening the back of the print with a clean, lightly dampened sponge or misting bottle. This expands the paper before mounting to match it to the substrate and help eliminate later paper buckling or possible warping. See Wet Mounting, page 40.

PRESSURE-SENSITIVES
The adhesive begins as a dry form, so the materials being fused during mounting should also begin dry. Take care not to store substrates in excessively damp or humid rooms, and if it is necessary, consider dehumidifiers or a rotation of your materials to promote time for them to dry out prior to mounting.

DRY MOUNTING: MECHANICAL PRESS
During dry mounting all the materials used in the mounting process should begin, bond, and remain dry...that means with no moisture. The only way to insure all the moisture is tapped from the mounting materials is through predrying.

All mounting materials, sans the adhesive itself, require the mandatory step of predrying when using a mechanical softbed or hardbed or press. Predrying taps water moisture from the materials before the project is exposed to any adhesive. Absorbent brown Kraft paper makes an ideal predrying envelope used in a closed, but not locked mechanical press only for 10-15 seconds.

Do not predry using a release paper envelope, as opposed to Kraft paper, because the moisture cannot be absorbed into the silicone and is trapped, sometimes turning to steam. When assembly line production creates numerous successive mountings needing to be cooled under a glass weight, use the same type of Kraft paper to absorb excess condensation from forming under the glass.

DRY MOUNTING: VACUUM PRESS
In opposition to the mandatory procedure of predrying when using a mechanical press, the principle suction of drawing the air from within the press precludes this step when using a hot/cold vacuum press. Begin each mounting day by running an empty press through one full, heated cycle to dry out all interior mounting materials. See Vacuum Press Maintenance, page 30.

> ***TIP:*** *GICLEES*
> The term *giclee'* a French term for 'spraying of ink' is used generally to denote high quality fine art limited edition digital images. Since they are often signed and numbered they should be preservationally mounted. TTPM will not apply to mounting of fine art giclees.

5

Wet Mounting

BASIC PRINCIPLES
Wet mounting is the oldest method of adhering paper to a backing board, using wet glues and pastes. It requires little financial investment and may be completed by hand with only a weight for pressure. It is a good alternative technique to hold in reserve for oversized items when equipment size limitations are an issue.

There are large production firms still strongly rooted in the practices of wet, spray and pressure-sensitive mounting. This proves that proper application coupled with extensive knowledge allows any mounting process to prevail.

The process, though economical, may be time consuming and messy. The permanency of successful wet mountings is often directly related to operator ability to properly apply an even layer of adhesive; allowing for appropriate dwell time; and adequate weight during drying.

TIME
Drying time is the time that an adhesive is allowed to dry in order to create a permanent bond. Anywhere from 3-24 hrs.

TEMPERATURE
Extremes of heat, humidity, or cold may lessen permanency of the bond. Follow all user recommendations.

PRESSURE
Plate glass or weights increase bonding during drying, but pressure from a cold vacuum frame is best. Pressure must be applied for a proper bond.

MOISTURE
If too much moisture is absorbed into the art or the adhesive is applied unevenly, materials will never bond well.

The largest variable when using wet glues involves the application of adhesive followed closely by the elements of time and pressure. Use of a cold vacuum frame will increase long term permanency by creating a stronger original bond.

Wet mounting may be less expensive, but the additional time involved during the application process, added to potential lowered permanency equals mounting costs with other options. All mounting processes end up costing about the same if operator time and permanence are calculated into the equation.

TIPS AND TECHNIQUE

A thick paste or liquid adhesive must be evenly applied to the print or substrate before positioning. The three elements of mounting to be controlled during wet mounting are time, pressure, and moisture. Temperature only becomes a notable issue when attempting to apply adhesives in an extremely hot, humid or cold environment, which might effect the flow, and/or drying time of the selected adhesive.

Basic mounting supplies include:
- a soft rubber roller 4-5" or semi-stiff brush
- commercial adhesive paste, liquid or conservation starch recipe
- piece of scrap glass
- 2 weights of ¼" plate glass or metal.

> OPTIONAL: Use of a cold vacuum frame will expedite the initial bond and overall longevity of the mount.

When selecting an adhesive, think through the reasoning behind using a wet bonding process. If minor corrections are required such as flattening folds in the print or replacing torn off pieces, repairs are easier to be made during wet mounting. It is a very controlled and safe way to handle and mount any item.

A wet paste should allow a fairly long open time and be repositionable. A safe paste selection would be nontoxic, nonstaining with age, and have long term bonding ability. Commercial pastes are available from major manufacturers that are starch based, neutral pH, nontoxic, buffered and water soluble for removal.

Conservation level wet pastes may be made as needed for special projects by mixing simple starch and distilled or sometimes deionized water. The addition of methylcellulose can give greater flexibility and a stronger bond to the mixture. Numerous recipes and cooking procedures are available from conservation sources.

APPLICATIONS AND PROCEDURES
STANDARD WET MOUNTING

1. Begin with a dollop of paste on a piece of glass, then roll the rubber brayer or foam roller across it to even out the adhesive. Using mat board scraps is not advisable for they will absorb the moisture from the paste accelerating drying and cutting down on working time.

2. Apply adhesive to the substrate rather than the print or photo. Stiffness of the mount board more easily tolerates the roller. Make certain the moist glue is evenly applied and covers every square inch of board.

3. Moisten back of the print by misting to expand the fibers to match that of the prepared wet substrate.

4. Align the print to the substrate across the top edge, gently sliding hand from the top to the bottom, first down the center then to the edges respectively to tack the print.

5. Check alignment to mount board, dwell time will allow for corrections if necessary.

6. Cover the print with a sheet of clean Kraft paper and gently spread it from the center to the edges to eliminate air bubbles. If the adhesive was applied to the substrate, exposed adhesive could stick to the Kraft paper; release paper may also be used.

7. Let project dry under weight for 4-24 hours, or optionally fuse in a cold vacuum frame, for the most permanent bond. Do not flex the project to reflatten until total drying time has been achieved.

BRUSH APPLICATION

1. If paste is thin enough to brush, apply to back of print working to achieve a smooth even coat of adhesive in a gridded pattern of both horizontal and vertical strokes.

2. Lay the print, face up, in proper position onto the selected substrate.

3. A dry 3-4" hake brush is used to smooth out the print and affix it to the substrate in preparation for weighting and drying.

This is closest to the traditional Asian methods of scroll mounting.

COUNTERMOUNTING

It is important to consider the problem of warping or bowing of the substrate when mounting any project, but this is particularly the case when wet mounting. The additional moisture applied to the art when misting or moistening it for fiber expansion, and the adhesive itself when applied to the substrate will both encourage the mount board to warp. Make certain to select boards thick enough to support both the size of the art and the added moisture. Thin X boards are generally suitable for 16x20" or less, 3X for larger than 16x20" and 3/16" foam for full sized 32x40" mounts. The thickness of substrate will depend largely on thickness of the art paper.

Even when all moisture problems are taken into account, the substrate is thick enough, the paper art is moistened for fiber expansion, and the project is dried under a weight for 24 hours, warping can still occur.

Warping is the result of the fibers on one side of the board becoming moistened during application of the glue, allowing them to expand. This expansion makes the surface larger than the drier paper surface on the verso side of the mount board, causing a natural curve. See page 93, Paper Grain and Fibers.

A dry board will lie flat when no expansion of paper fibers has occurred through the introduction of moisture.

Glue side

The paper fibers on the top have expanded with applied moisture, while the bottom of the board is holding at the original size. The two alternate lengths create a bowing or warping of the substrate.

The only way to compensate for the fiber expansion on the surface of the mount board is to apply the same degree of tension to the back of the project. By wet mounting a sheet of approximately same weight paper to the back of the board using the same adhesive and technique the surface tension will be the same both front and back, reducing or eliminating the bow.

WEIGHTING VARIATIONS

After the art has been aligned, burnished, or rolled into position, it must be weighted during the drying process. Lay the substrate and artwork face up on sheet of ¼" plate glass. Layer a sheet of spun nylon (Pelon) then a blotter on top of the art, then place the second sheet of ¼" plate glass. The Pelon prevents the adhesive from sticking to the blotter, while the blotter absorbs wet adhesive moisture. The glass is both cool and heavy. Dry blotters should replace damp ones every hour or two for the first few hours, then morning and night the next day.

OVERSIZED POSTER MOUNTING

Larger format wet mounting is made easier to handle by counter rolling the art face in for initial positioning. Pay attention to the required open time for the wet adhesive to activate properly for bond.

1. Apply adhesive with rubber roller or brush as indicated in "standard wet mounting" to substrate creating an even double coat.

2. A large flat poster is easier to handle if loosely rolled just before mounting, then align the top edge and unroll onto wet substrate.

3. Cover with clean Kraft or release paper and roll from center to outer edges.

4. Place under a weight to dry or in a cold vacuum frame for 3-12 minutes to accelerate drying through the vacuum action. Adhesive is better forced into all the paper fibers and variations for best wet bonding.

FLOATED ART

Dollar bills, doilies, even papercut art may be floated onto glass or an opaque substrate to create the visual effect of floating within the frame. Tiny dots of acid-free PVA (white glue) are perfect for this technique. Apply the dots as small as possible while still being enough to hold.

PAPERCUT ART

Papercutting is one of the oldest and most enduring art (craft) activities in the world. In China it is known as *chien chih*, with the earliest samples from the Han dynasty (206 BC to AD 220), when gold and silver foils were used instead of paper. In Japan papercutting is called *monkiri* and it produced decorative lightweight colorful possibilities for decoration and storytelling. In Germany and Switzerland the art is known as *scherenschnitte*. In Poland it is called *wycinanki*, and The Dutch call it *knippen*. Many cultures have embraced this traditional form in one way or another. In Jewish cultures it is part of their religious artwork and holiday celebrations.

These highly detailed lightweight patterns are extremely delicate and easily damaged during framing. They come as single or multiple layered papers and have a natural three dimensional quality. Sprays may only be applied by spraying above the art letting the fine mist settle gently on the back of the cutting. A wet mount option is applying tiny small dots of acid-free PVA. Depending upon the weight of the cut paper and the detail, the small dots should be as small as ¼ the head of a small straight pin. A toothpick tip full of glue is excessive and might saturate through the art.

Papercut art is produced with highly light fugitive papers that will fade even when framed with UV glazing. Dye based papers are faded by visible light as well as UV light.

ENCAPSULATION

Conservation encapsulation is encasing an item between two sheets of clear plastic. Polyester, polypropylene, and polyethylene. They must be pure without additives to be used for encapsulation. Polyester film Mylar-D (Dupont-Teijin), its European equivalent Melinex 516 (Melinex 455 or 456), or Hostaphan 43SM all meet long term Library of Congress storage requirements, are clear, strong, smooth and rigid. Cut sheets about 1" larger around the item and seal all four sides with ½" wide 3M 415 or 889 polyester double-sided tape. It may then be floated in a recto verso (two sided window) mat or mounted to a collaged background without damage.

This is a conservation mounting alternative to wet glue and is often the best choice solution to wet or pressure-sensitive adhesives for delicates, thermographics, and items if uncertain origin. Encapsulation of papercut art would be the best alternative. It is archival and will better protect delicate art. Encapsulation also works for wrapping and mounting magazines, books, letters and other bulky and sometimes heavy items.

BLUEPRINTS (a.k.a. CYANOTYPES)

These are architectural drawings often made by a thermographic process. Depending on their age they can be damaged by dry mounting, so alternative encapsulation or P-S methods should be considered. Contemporary blueprints are diazo developed with ammonia fumes and are highly susceptible to light and chemical damage. They are unstable in an alkaline (calcium carbonate buffered) environment which could accelerate their deterioration by turning them brown. They should be encapsulated or preservation hinged and matted with unbuffered neutral pH 100% cotton rag photo board, under UV glazing. Copies should be made so the original may be held in dark storage.

> ➤ **TIP:** *DIGITAL IMAGES*
> Since many digitals images are printed with water based inks, use of wet and spray adhesives should be used carefully and only when familiar with the process.

> ➤ **TIP:** *ALCOHOL CLEANING*
> Some wet glues resist bonding to nonporous materials that have surface coating, like RC photographs. Clean the back of the photo with alcohol or adhesive solvent to remove developer residue or oily fingerprints that might prevent a good bond. Then apply adhesive to photo rather than substrate and place in cold press 10-12 minutes.

6

Spray Mounting

BASIC PRINCIPLES
Spray mounting is another economical mounting alternative requiring no equipment, though a spray booth or ventilation hood is highly recommended.

Permanence of a spray adhesive is directly effected by various factors. Choosing the right adhesive; applying an adequate amount; selecting the appropriate substrate; and using correct techniques such as pressure during drying, all contribute to good bonding. Longevity of sprays also depends upon the dryness or humidity level in the materials.

Spray manufacturers define permanence as *tear strength*. By that standard, properly applied permanent sprays will indeed tear paper fibers before releasing the bond demonstrating impeccable permanence. Longevity vs. tear strength remains the argument. Proper application procedures ensure successful mountings and the bond remains as long as tackifiers exist between art and substrate. Once dried out all long-term permanence is diminished.

TIME
Open time is the window of time in which the spray-coated item can be mounted (approx. 3-10 min.) Some prints may be repositioned during open time. *Bond time* is the time required for a permanent bond to take place.

TEMPERATURE
Most manufacturers have a temperature range which spray adhesives will withstand.

PRESSURE
Pressure is required to create the necessary contact between adhesive and substrate. A vacuum frame is recommended for maximum pressure during drying.

MOISTURE
Condition the art and substrate to the same environment before mounting.

Selecting the correct spray adhesive for a job is vital. Check for porosity and permanence levels. The correct substrate for each individual project also effects long term bonding. Sprays resist bonding to oil-impregnated surfaces or wax-coated materials. Substrates including hardboards, fiberboard, MDF, and low-grade newsboard have oils that can soften the adhesive, resulting in bond failure.

TIPS AND TECHNIQUES

Spray adhesives are available as solvent-based and water-based. Appropriate safety measures are necessary when using traditional solvent based spray adhesives. Carefully read all the information concerning aerosol adhesives on the label before using them. Proper ventilation during use is required. Adequate ventilation defines as the flow of enough air to dilute contaminant, either vapor or dust, to a safe level. An open window or door is adequate for occasional aerosol use but regular spraying requires a spray booth and/or exhaust fan to meet safety standards.

If building a large walk-in spray booth is necessary, make it large enough to adequately contain all overspray and accommodate all sized projects. A formal booth may need to meet local permit, electrical and environmental standards, or regulations. Wearing a breathing mask for added protection is also advised.

Good TTPM technique is important. The three basic elements to address with spray mounting are time (both open and bond time), pressure and moisture. Adequate time after applying the spray is required for the solvent to evaporate and the adhesive to become tacky. Solvents are what allow the adhesives to be sprayed in the first place, but evaporation of excess solvents must occur to initiate bonding. This is when the print may be aligned and repositioned if necessary, and is known as *open time*. Though the initial bond is made within the first hour, maximum bond is achieved in 24 hours.

A rubber roller, squeegee, or vacuum frame may be used to create the necessary pressure during hand bonding. Glass or metal weights are required for pressure when a vacuum is not available. A cold mounting vacuum frame is recommended because it creates a more intense, even pressure, speeds drying and creates the bond in less time.

Pressure cannot be stressed enough. If mounting by hand, place the project beneath a glass or metal weight 12-24 hours during bonding. In a cold vacuum frame (see Chapter 8), maximum bond occurs in 1-5 minutes. Nonporous RC photos often require additional time (up to 12 minutes). In a heat application, bonding takes place in 1-2 minutes at 175-195°F in a mechanical softbed press.

Basic spray mounting materials and supplies include:
- Aerosol spray adhesive
- Spray booth or ventilation hood
- Accordion support
- Kraft paper
- 4-5" Soft rubber roller or squeegee
- Weights, glass plates or cold vacuum frame

Accordion base for project support

Scored mount or mat board makes a good
support, as it will better hold its shape
if saturated with spray adhesives.

Portable Screen with Hinged Hood

The art is supported on the tops arches of the
accordion to keep it from accumulated adhesive.

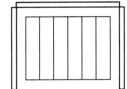

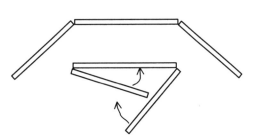

This portable spray booth may be easily
folded for storage when not in use,
but extremes of overspray may encourage the
folding leaves to stick if still wet when stored.

Free Standing Booth

Excess use of spray adhesives will require
appropriate ventilation facilities that meet all
OSHA health standards.

A spray booth permanently set-up with a ventilation
hood, fans, and exhaust system.

ADHESIVE TYPES

The are many good sprays on the market. Check their flexibility with both porous and nonporous items. Since aerosol spray adhesives are technique sensitive, the speed of application; distance from substrate during application; and open time; all influence bond strength. Select the correct adhesive for each individual job. Multi-purpose aerosols offer extra-high coverage and will hold almost any material, though the Library of Congress does not recommended sprays for RC photographs. The adhesive is fast drying, resists humidity and will not soak through most surfaces or wrinkle thin mountings.

Sprays are suggested for use with fabrics, paper, and boards and bond without staining, yellowing, or cracking. They are safe for use with heat-sensitive items. Make certain to select the spray that fits the specific needs of the project, and check product label for positioning flexibility (repositioning) during the 6-8 minute open time.

Some popular sprays include: 3M Spray 77 mounts with 15-30 minutes open time, lays on the surface without soaking in, and may be applied to one or both surfaces for extra bond. 3M Vac-U-Mount has a 2-6 minute open time with a 10 minute maximum for placement and mounts in a cold vacuum frame in 1-2 minutes. Spraytex Good Glue Spray is for both porous and nonporous art, mounts by hand or in a cold frame in 2-5 minutes, or in a hot mechanical press at 175-195°F in 1-2 minutes. Sure Mount Spray is an acid-free, water-based, and reversible starch for paper and fabric that mounts by hand in 3-12 hours or in a cold frame in 2-5 minutes, and hot at 190°F for 1-4 minutes. Nonporous photos could take up to 10-15 minutes in a heat press.

> ➤ *TIP: HEATING SPRAYS FOR REMOVAL*
> As with wet glues, removal of dried out sprays may sometimes be hastened by heating in a dry mount press or using a hand-held shrink-wrap gun or hair dryer.

> ➤ *TIP: WRINKLE REMOVAL*
> Flatten wrinkled, creased or curled paper by reverse rolling with a blotter cover and placing in a mailing tube for a few hours. If additional flattening is required, layer a sheet of Kraft paper over the art, mist with water and place in vacuum frame. After water is visibly removed from the Kraft paper (approx. 5 min.) immediately apply adhesive to the board and mount it. The art may feel slightly damp but remaining moisture will be removed during the vacuum process.

APPLICATIONS AND PROCEDURES
STANDARD SPRAY MOUNTING
Use only in well-ventilated area, keep away from flames and minimize overspray by applying at a 90-degree angle only 6-8" from work.

1. Read label and point arrow on nozzle to black dot on can to indicate proper alignment. Shake can to mix thoroughly. Test spray away from art to insure proper flow and clean application.

2. Begin spraying off the edge of the print or substrate to prevent puddles and globs of adhesive, and extend beyond the opposite edge before stopping. Slightly overlap wet bands of spray. Apply an adequate amount of adhesive spray, first horizontally then rotate board ¼ turn, and repeat. Adhesive may be applied to either substrate or print, but be careful not to contaminate print front with overspray.

Apply spray then rotate substrate 90-degrees

Begin off the left edge and continue past the right. This may be done in one continual motion or in separate left to right passes across the substrate.

3. Allow for appropriate open time (3-10 minutes, read product directions) while solvents evaporate and adhesive becomes tacky.

4. Position print, cover with Kraft paper and smooth from center to edges using flat of hand or soft rubber roller.

5. Lie flat under weight, allow time to dry and bond. Placing in a cold vacuum frame will expedite the mounting and produce a bond much quicker.

6. Invert can and clear nozzle of remaining spray after use.

> ***TIP:** SELF-SHAPING*
> The fine delicate paper cut areas of Japanese and European cut paper designs will NOT tolerate the aggressiveness of separation for self-shaping of a pressure-sensitive film.

JAPANESE PAPERCUTS

Delicate problem mountings are well suited to spray mounting techniques. The intricacies of Japanese papercuts with their multiple openings makes wet, pressure-sensitive and dry mounting all unacceptable due to adhesive being visible between the cut outs and damage potential.

Do not spray directly onto the art, it is delicate and could be damaged. Great care needs to be taken when applying spray so not to puddle it or run it under the cuts to the front of the art, it is best to mist the adhesive from above. The spray will settle gently down onto the papercut which should be supported by an accordion piece to keep the art out of the adhesive. Be very careful when handling this type of artwork, and respect its delicate nature. It is so fragile that even the draw of a cold vacuum or the clamped closing of a mechanical press can tear the sections.

1. Size and prepare materials. Select a spray that will reactivate and mount with the application of heat, such as Vac-U-Mount, Good Glue Spray or Sure Mount Spray.

2. Lay paper art facing down onto an accordion fold small enough to support all loose cutout portions. This will keep the artwork out of the puddles and support it. A flat sheet of release paper may also be used but not Kraft paper.

3. Shake can, test spray, check label for appropriate distance to be sprayed from art.

4. Spray above art allowing adhesive mist to gently float down onto cutout. Do not apply spray directly to cutout. Overspray can easily bleed under the delicate free form pieces. Every part of the cutout must have adhesive applied for adequate bond.

5. Invert can, spray to clear nozzle.

6. Let sit until **completely dry** or no longer tacky to the touch.

7. Place into position on matboard backing, cover with release paper, and press with a small tacking iron to reactivate the spray and hold it in place.

8. DO NOT place in a cold vacuum, hot vacuum, or hot mechanical press. Even the slightest shift of air during compression of the press can damage the delicate thin papercuts.

These are very light fugitive pieces and often fade in the visible light that reflects their own color as well as UV light.

7

Pressure-Sensitive Mounting

BASIC PRINCIPLES
A pressure-sensitive adhesive (PSA or P-S) is a permanently tacky substance that bonds to an untreated surface at room temperature, with only the application of slight pressure. By definition P-S adhesives differ from every other category of adhesive.

PSAs do not require any open time or solvent evaporation in preparation for bonding. They may have slightly lower bonds than heat-set adhesives, but because of their permanent tack, they are always ready to bond to almost any surface with only thumb pressure.

They are dry, synthetic adhesives that are clean, easy to use, odorless and require no heat or solvents. These adhesives have release papers applied to protect them from bonding until required to do so. There are a number of commercial pressure-sensitive products on the market, available as adhesive sheets, rolls, and preadhesived boards with release paper coverings. PSAs may be either permanent or removable and are available as low, medium and high tack varieties.

TIME
Maximum bond strength is not achieved until the adhesive has set for 24 hrs.

TEMPERATURE
Though not a dominant element, the warmer the materials, the more aggressive the bond. After initial bonding, extremes of heat and cold can effect the long-term bond.

PRESSURE
A weight or vacuum is required to supply the pressure to create a permanent bond.

MOISTURE
Damp materials will not bond together regardless of mounting methods. Make certain all materials are dry.

A permanent PSA means it is not possible to remove it without damaging the substrate, such as a mailing label. A removable P-S may be removed without harming the substrate in any way or leaving adhesive residue, like a post-it-note.

Ease of removability depends on the selected substrate, length of time it contacts the substrate, and temperature during bonding. If surfaces are dirty, oily, or too textured they will not bond as well as to a smooth surface. The warmer the materials, the more aggressive the adhesive becomes.

High tack PSA is aggressive and most difficult to hand apply since it has no repositioning potential. It is applied hand guided with roller machines and has been traditionally geared to the commercial market, until digitals. Medium tack PSA is used in framing. It may be repositionable until burnished in place. Low tack PSA is not suitable for the longevity of framing demands. It may be found as a temporary positioning tack used in preparation for other bonding, as with HA adhesives.

P-S FILMS: Films are available with and without carriers, both having positive applications. 3M Positionable Mounting Adhesive (PMA) is a sheet of low to medium tack 100% adhesive with release paper. Having no center carrier, the adhesive is rather delicate and easily pulls apart by handling. It is capable of self-shaping for odd shaped projects and small prints, perfect for montage and collage applications. Self-shaping is the process of the adhesive separating to the shape of the artwork without additional cutting or trimming. PMA is not recommended by the manufacturer for mountings larger than 20x24".

Crescent Perfect Mount Self-Adhesive is a medium tack pressure-sensitive film with release liners. This film consists of a very thin polymer sheet coated on both sides with pressure-sensitive adhesive. This allows the sheet to be handled more aggressively without pulling apart, though it will stick to your fingers and could wrinkle. It cannot shape to the artwork as the PMA film with no supporting carrier, and requires trimming to shape. It is neutral pH and trims cleanly with a blade without pulling. It is repositionable until burnished in place and requires weighting while creating a permanent bond. Neschen Gudy 870 is an aggressive nonrepositionable high tack P-S film with excellent long term bonding capabilities.

P-S BOARDS: Pressure-sensitive boards have the film adhesive already applied with a protective release paper that is removed before mounting. They are available as low tack repositionable and the more permanent "peel and stick" versions in medium and high tack. Although some are considered acid-free, always remember that permanently attaching anything to a backing board is <u>NEVER archival</u>! Some boards are neutral pH and acid-free. Check specifications for P-S products from manufacturers including Nucor, International Paper, Crescent, 3M, Hart, Nielsen Bainbridge, and Bienfang.

No major equipment purchase is necessary for pressure-sensitive mounting though roller machines or cold vacuum frames are recommended for greater permanence. The 20" Schild C-35 PMA Applicator will handle substrates up to ¼" thick. Heat presses may be used to accelerate the initial bond, though that defeats the purpose of a cold application.

P-S TAPES: Tapes should never be placed in direct contact with artwork or photograph, regardless of whether they are acid-free or not. Damage can occur from all P-S adhesives over time. Even water soluble brands will only remove the tissue carrier and not the adhesive saturation that has occurred. Both self adhesive and PVA gummed paper tape will chemically react and cross-link with atmospheric pollutants after about ten years rendering the new substance (adhesive) insoluble, and nonremovable.

TIPS AND TECHNIQUES

There are two dominant elements to be controlled when using pressure-sensitives, time and pressure, but warm materials bond more aggressively than cool ones. Although pressure-sensitive applications appear quite well bonded almost immediately, their true ultimate bond will occur only after 24 hours. Heat may be applied at 190°F for 1-3 minutes to expedite maximum bond.

Remove the release paper, apply artwork, cover, and burnish down to create the bond. Basic supplies required are minimal:
- Adhesive film or pressure-sensitive board
- Release paper or Kraft paper overlay
- Squeegee, or rubber roller
- OPTIONAL roller press or cold vacuum frame

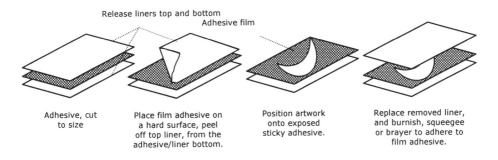

Remove top liner, trim to shape, and place into position onto selected substrate, cover with liner and burnish into final mounting place.

P-S mountings (artwork image) should be matched to substrates so the expansion and contraction will be equal. These adhesives will creep, then hold on during contraction allowing for bubbles and creases, particularly on large images. A substrate and image should expand and contract in unison to avoid blisters and separation.

APPLICATIONS AND PROCEDURES
PRESSURE-SENSITIVE FILMS
Films are a quick and easy way to apply a clear pressure-sensitive adhesive to a project with little mess and no major equipment required. Especially applicable to use when control over the substrate color is required, as to control ghosting or to capitalize on color tinting.

STANDARD FILM APPLICATION
1. Cut film with liners intact about 1" larger than desired project.

2. Cut all remaining materials and boards to size.

3. Lay release coated film on a flat surface and gently roll liner from the top (face up side) of the adhesive. While still lying flat, apply art to the newly exposed sticky adhesive film.

4. Cover adhesive and applied art with the removed release liner and burnish the art to the adhesive through the liner with a burnisher, rubber roller, squeegee, or run through a roller machine to bond first side.

5. Remove protective release liner cover and trim project to size, cutting away excess exposed P-S adhesive.

6. Invert trimmed art face down, peel opposite liner from the verso of the art by rolling it gently off from one end to the other. This prevents the art from creasing or the film adhesive from letting go during preliminary application.

7. Position adhesive backed art project onto desired substrate, cover with release liner and burnish into place. Apply setting pressure from center to outside edges to remove any potential air from beneath mounting.

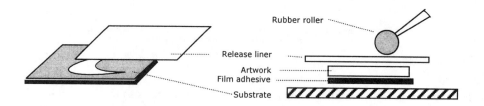

8. Set under weight for 24 hours, or run through roller press or cold frame.

SELF-SHAPING

A technique limited to PMA pressure-sensitive film without a carrier allows for items of any shape to be premounted to the pure film, then peeled from the remaining release liner to be positioned for final bonding. The adhesive will separate from the liner in the shape of the desired artwork without the need for trimming to size or shape. Some clean up may be needed around the edges of the art. Stretched adhesive has the tendency to show up along the front edge after stretching during separation. There is some degree of resistance during the separation of the adhesive from itself, making this process unacceptable for delicate or fine detailed self-shaping such as Japanese paper prints.

P-S films with carriers do not have the ability to separate to the shape of the removed art because the carrier is a solid sheet. All shaping must be manual with a blade or scissors.

GHOSTING NEWSPRINT

Ghosting is the undesired bleeding through of text or pictures from the verso side of a mounting. The most commonly framed source of ghosting comes from newspaper clippings or magazine articles. The paper used for printing is inexpensive, porous, and thin enough to readily see the printing from the back side through to the front after mounting.

Pressure-sensitive films are a good choice for speed and simplicity of application and will not soak into porous papers. Since there is writing on the back, a dark substrate is required to camouflage the verso text. This eliminates use of white pressure-sensitive boards, but films being clear are perfect for use on any colored substrate. Follow procedure above for standard film applications.

COLOR TINTING

Though ghosting is something to avoid for its negative appearance, color tinting is the positive use of a brightly colored substrate to better enhance or draw a color out of a given project as a decorative design element.

- Thin rice papers may be mounted to surface mat boards to create new designer textured mats.
- Sheer solid color fabrics, such as thin silks and linens, may be tinted with pastel hints for shadow boxes.
- White certificates and stationery may be tinted by the substrate to better match the colors of a room.

This design application of mounting is achieved by using a clear adhesive on a colored substrate or mat board to enhance any thin, sheer, neutral colored material. Any time a color tint is part of a design additional charges must be applied for the additional design concept, regardless if the chosen colored substrate is a higher priced substrate or not.

FLOAT OR PLAIN MOUNTING
This technique is used for centering or randomly placing a print or photo to a colored mount board as an option to matting. Often used in photography and photo competitions, it is used by first applying adhesive to the photo, trimming to size and shape, then mounting the preadhesived project to the selected substrate. See chapter 10.

FLUSH MOUNTING
This method of application allows for placement of the project to a selected substrate without centering. After appropriate bonding time has passed, the art and substrate may be trimmed at the print edge allowing the exposed board or foam to suffice for a frame trim. This might also be called bleeding the artwork to the edge of the board. The edge may be straight, bevel, or reverse beveled for a finished look.

After bonding, flush mounted substrates may be blunt cut, beveled or reverse-beveled to finish the edges.

See chapter 10, Dry Mounting, for more explanation on these finishing techniques.

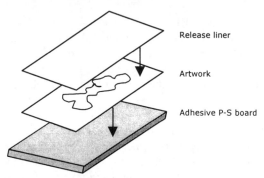

Release liner

Artwork

Adhesive P-S board

> **TIP: CREEPING P-S ADHESIVE**
> A pressure-sensitive adhesive can ooze or creep with high temperatures and humidity extremes. Both substrate and image expansion should be matched (paper to paper, photo to foam) to better prevent bubbling (bond failure) during uneven contraction of the two surfaces.

> **TIP: FACTS – MOUNTING GUIDELINES**
> Mat and backing board standards have been introduced by FACTS for maximum preservation framing. Though mounting and laminating is not preservation, the materials selected should still be maximum quality for best long term bonding results. See ANSI/ISO 39.481992; FRM-2000 4.05, 6.02, 7.02; and PMMB-2000 at www.artfacts.org for guidelines.

PRESSURE-SENSITIVE BOARDS
STANDARD BOARD APPLICATION

When the color of the substrate does not matter, pressure-sensitive boards may be used. They are available as mount board and foam, in varying thicknesses. They are easy to use and require no additional equipment besides a weight, though rollers or cold vacuum will improve the initial bond.

Repositionable PSAs may be more aggressive with nonporous photos than with porous papers. To test the nature and degree of repositionability, lightly touch a small corner of the art to the exposed P-S board before aligning.

REPOSITIONABLE ADHESIVE
1. Size board to fit paper art.

2. Remove liner by peeling from the corner, separating it from the sticky adhesive and board. Do not touch adhesive with fingers. Oils could restrict proper bond.

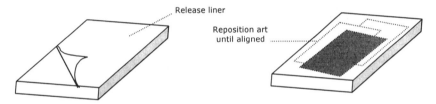

3. If repositionable, check for proper alignment along top edge or center and for trapped air bubbles, remove and reposition if necessary.

4. Layer the release sheet on top or use Kraft paper to protect print, reinforce bond by rubbing with hand or soft roller from the center to the outer edges.

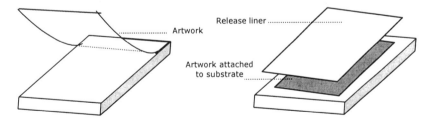

5. Weight for up to 24 hours to allow adhesive to cure, run through roller or place in vacuum frame to expedite bonding.

NONREPOSITIONABLE/PERMANENT ADHESIVE
1. Size boards and materials as needed.

2. Nonrepositionable boards grab rather aggressively when touched by any mounting. Peel back top few inches of release liner to expose adhesive board and crease flat, or fully remove release paper backing to break the tight bond, then temporarily realign onto board, exposing top 2-4" and of exposed adhesive.

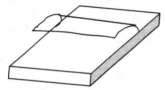 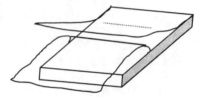

3. Position artwork by squaring the top edge of board with the edge of the artwork.

4. Carefully peel liner from board out from under the art as smoothing the art flatly into position.

5. Cover with release liner, burnish and weight to cure.

LIBRARY OF CONGRESS PHOTOGRAPH RECOMMENDATIONS
By using pressure-sensitive films, rather than P-S boards, a more rigid board designed specifically for use with chromogenic photographs may be selected as the substrate. Numerous mount and mat boards have been developed by Crescent, Bainbridge and other major manufacturers to meet Library of Congress standards for archival photo applications as designated below. New 1998 American National Standards Institute (ANSI) standards for black and white photos specify a pH range of 7.0 – 9.5 with a 2% calcium carbonate reserve. Color and diazo materials should have a pH range of 7.0 – 8.0 with a calcium carbonate reserve of less than 2%. They should be 4-ply boards of 100% cotton fibers.

In addition, the boards must meet the following specifications by being:
- Rigid enough to support their own weight leaning against wall
- Strength to self-support when held with two hands at one edge, without extensive bow
- Surface smoothness
- Clean cutting density
- Resistance to impact without breaking
- Free from warpage

RC PHOTOGRAPHS
Orange peel is a common problem with RC photographs when mounting. This is the lumpy visual contouring of the softer photo surface emulsion to the contour of the selected substrate. Smoother substrates create less orange peel. Although most detectable during dry mounting, even aggressive hand pressure during pressure-sensitive mounting will create this effect.

ILFOCHROME CLASSICS (CIBACHROME)
The difference between RC and Ilfochrome Classic is simple. RC photos are resin coated paper front and back with plain paper in the center. A Cibachrome is 100% polyester, and has no paper core. Neither type of photo is porous. A Cibachrome is greatly changed when exposed to heat whether or not an adhesive is under it. The polyester base relaxes to follow the contours of the substrate almost always creating orange peel. The best mounting solution would be hinging, followed by pressure-sensitives which best maintain the dignity of the original image.

Ilford also manufactures <u>RC Cibachrome</u>. Though called a Cibachrome it consists of a paper core with resin-coating on either side just as any other RC photo. It may be mounted and otherwise treated like any other RC photo and often has Ilford Cibachrome printed on the backing paper.

 TIP: *HEAT DAMAGE*
High temperatures of 200°F may damage the gloss emulsion creating small blotchy areas that look like scuffing or snowflakes. Extended exposure to high temperatures may also dry out and crack surface emulsions.

BRASS RUBBINGS
Since the papers used for rubbings are lightweight, P-S film and boards are good mounting options. The lightweight paper expands when wet or spray mounted as part of the process. Because wax is not water soluble the paper beneath it cannot expand with moisture and may cockle around the perimeter of the wax making it difficult to flatten. The waxes used to create brass rubbings have a much higher melting point than most dry mount tissues making dry mounting possible, particularly when using 150°F SpeedMount HA foam board from Bainbridge.

DIGITAL IMAGES
Since digital images are generally water based inks on lightweight paper stock pressure-sensitive mounting is the best choice. Low temperature HA dry mount boards (ie: Bainbridge *Speed*Mount) may also work with some images, see Appendix, pages 145-151 for additional suggestions. Signed and numbered giclees are to be preservation mounted only.

8

Cold Mounting

BASIC PRINCIPLES
Anytime a vacuum frame or roller laminator is used with a wet glue, spray or pressure-sensitive adhesive it is generally referred to as cold mounting. All of these adhesives may also be applied by hand methods using weights as noted in the previous three chapters. Cold mounting thus defines as any mounting technique requiring no heat.

Cold vacuum mounting is a mechanical system using a combination of both adhesive and vacuum suction to create the bond. Wet and spray glues may be used in this process as well as some pressure-sensitive adhesives. Almost any artwork, including heat-sensitive items, will tolerate cold vacuum mounting. *In the framing industry,* any mounting without heat is considered cold mounting.

In the photo industry, cold mounting generally refers to mounting without heat using pressure-sensitive adhesives or pressure-sensitive laminates. A thin sheet of adhesive film, or film with carrier, may be hand or machine applied. Small hand applications are kept under 16x20" in size when used with a rubber roller or squeegee. Larger applications may be made with double roller machines, called roller laminators, designed for large-scale operations. Roller adhesives come with single or double-sided release liners and are high tack geared for commercial applications, such as mounting transparencies for light boxes, and 4'x8' ad campaigns, and wide format digitals.

TIME
Average vacuum frame times are 4-15 minutes long.

TEMPERATURE
There is no temperature requirement besides average room temperature, and advised manufacturers' guidelines for individual adhesives.

PRESSURE
Correct pressure is controlled by the vacuum frame itself, which varies automatically to accommodate different substrate thicknesses.

MOISTURE
The vacuum draws moisture automatically from the project as the air is sucked out.

COLD LAMINATING
Lamination is the process of applying a durable, clear plastic film to a flat surface for the purposes of protection and enhancement. Some items that may be sensitive to heat-seal laminates would be ideal candidates for cold laminating processes. Originally used for ID cards and menus, use of cold plastic laminates has been in commercial photography since the 1980s. Today they are used in sign and display for indoor and outdoor applications, floor graphics, and for assorted image protection. Photographs, maps, brass rubbings, and thermographic papers can be easily laminated with roller equipment and cold methods.

Roller laminators are traditionally large commercial double roller machines designed for use cold, or with one or both rollers heated for adhesive and/or laminate application. They range in size from 14" to 80" wide having been used dominantly in production photo labs, advertising agencies, and commercial graphics houses. PVC (vinyl), polyester, and polypropylene over-laminates and encapsulates are all used for surface protection. The digital print market has now made cold rollers a viable option in custom framing too.

Adhesives are extremely high tack and are made up of a wide variety of acrylic pressure-sensitives. Their highly aggressive nature recommends against using them in a cold vacuum frame because of alignment difficulty air removal. Roller machines operate somewhat like an old fashioned wringer washer, but unlike office heat-seal encapsulators, are capable of mounting to boards up to 1" thick. Laminates are available in various polyester and vinyl thicknesses, with numerous finishes and textures, new option for framers.

COLD FACE MOUNTING
Face mounting methods use either a wet silicone system of application or a dry clear P-S roll adhesive and roller laminator to mount display materials that require permanent durable protection. Face mounting may be used for ordinary prints or translucent materials bonded behind glass, acrylic sheet, polystyrene, or polycarbonate.

This process is one of the current methods for exhibiting display photographs and digitals in some of the major museums. It is used for Cibachrome photographs in advertising, and polyester film to acrylic sheet for lightboxes. It is a permanent process and face mounting should not to be attempted by a custom framer untrained in this technique.

MOUNTING POLYESTER ENCAPSULATED CHARTS
Two-sided encapsulated items are commercially produced using roller laminators with variable thicknesses of laminates. Since polyester will not absorb adhesive moisture, there is no wet; spray or dry mount adhesive aggressive enough to hold slick polyester encapsulates to a substrate. Sanding or scuffing the back will still not allow the adhesive to penetrate to have anything to bond to. Only a truly aggressive, high tack pressure-sensitive applied with a high-pressure roller machine will hold the laminate in place without bubbling for framing. Locate the nearest large commercial production lab and subcontract the project to them for mounting.

VACUUM FRAME BASICS

Cold vacuum frames, combination hot/cold vacuum presses, and mechanical dry mount presses, make up the three basic types of mounting equipment systems currently on the market. The *cold vacuum frame* uses vacuum mounting action with spray, wet or pressure-sensitive adhesives in conjunction with pressure, but no heat. It is often considered more convenient, efficient and less fearful than dry mounting. Since cold frames offer no option for heat, they are unable to tap into the laminating and creative possibilities that heat vacuum systems offer.

Since vacuum frames do not use heat, only three of the mounting elements are involved. Time may vary with the individual adhesive selected and the overall size of the project. Open time of adhesives can vary from between 3-15 minutes, allowing for repositioning. Temperature remains at room temperature for cold mounting. The rubber bladder, or diaphragm, being pressed up against the mounting project controls pressure, and moisture is drawn from the unit as the vacuum is pulled. Mountings may not be achieved if the project is larger than the press itself. The seal along the edge of a vacuum press must not be invaded, for if the seal is broken the vacuum may be lessened or eliminated effecting the pressure required for a good mounting.

Use a sheet of Kraft or release paper to protect the glass top from adhesive contamination. All vacuum mounted artwork should be left in the shop overnight to check for bonding permanency.

FLATTENING ARTWORK

Wrinkled and buckled artwork may be somewhat flattened by dampening a sheet of Kraft overlay, then placing it into a cold vacuum to straightened the moist fibers and draw the moisture from the art.

1. Lay artwork face down on a clean surface, carrier, or mountboard.

2. Cover it with heavy Kraft paper and spray the paper with water so it absorbs the moisture.

3. Spread a towel over the dampened paper to absorb excess water and evenly coat the paper.

4. After the water has adequately dampened the art, place all three (board, art, damp paper) in the press for 3-5 minutes.

NOTE: This is a somewhat antiquated method of flattening with the more noninvasive moist blotter/Pelon method or humidifier being the preferred version. See Appendix, page 140.

HEAT-SENSITIVE ITEMS
It is sometimes difficult to determine heat sensitivity, but a slight touch to the edge of a tacking iron will immediately darken sensitive thermographic papers. Raised lettering, 4-color copies, faxes and even RC photos are considered to some extent heat-sensitive. Test items for tolerances prior to mounting to prevent damage. In the 21st century the plot thickens with digitals and digital photos, some being sensitive some not.

THERMOGRAPHY
THERMOGRAPHIC PRINTING
In printing, thermography, meaning *heated printing*, is a raised-letter printing process that simulates engraved printing by applying a fusible powder to an oily, slow-drying wet ink prior to drying. The substrate is then passed under a heater to fuse the ink and powder. Some inks contain an expansion agent that, when exposed to heat, creates a raised letter effect Cold mounting methods of wet, spray or pressure-sensitive applications are suggested.

THERMOGRAPHIC PHOTOCOPYING
In thermographic photocopying or *writing with heat*, an original document and sheet of special copy paper are passed together through a machine that exposes them to heat, or infrared rays. The ink used on the original must contain a metallic or carbon compound in order to absorb the infrared radiation. The copy paper contains a heat-sensitive substance fused between a transparent sheet of paper and a white, waxy backing. The original heats the darkened areas of print and transfers the image to the copy paper. Thermographic copy prints must be cold mounted.

IMPACT VS. THERMAL NONIMPACT PRINTERS
Impact printers and dot matrix printers use an inked ribbon to pressure apply an image onto a sheet of plain paper. Nonimpact printers require thermal or electrostatic transfers of the original image rather than mechanical means as above. Bubblejet and inkjet printers squirt heated ink through a matrix of holes to duplicate the copy images. Laser printers form the image on a selenium-coated drum using a laser beam that transfers the exposed image from the drum to the paper. Thermal wax transfer printers and dye sublimation printers use heat to transfer color dye or pigment from ribbons to special papers.

FACSIMILE (FAXES)
Derived from the Latin fac simile *made like* it refers to a process of reproducing graphic material from a distance. A light sensitive device scans the image to produce an electric signal that is sent over telephone lines. Reproduction of the received copy can be on photographic, electric-sensitive, or thermal (heat-sensitive) paper. Any fax produced from heat will continue to darken when further exposed to any heat source. Warmth from the sun, a mounting press, or even heat inside a frame may activate the paper darkening whites and blackening grays of thermographically produced faxes.

DIGITAL IMAGING

Trends in technology and art greatly impact the framing industry. The once disposable digitally printed proof (prerequisite for the lithograph) has evolved into fine art as the *giclee'*. First the disposable proof, then four color copiers, one-hour developing laboratories, and home office inkjet printers. Fine art giclees are to be treated as any other signed limited edition utilizing preservation mounting techniques. All other digitally created work should be framed according to their monetary value, emotional value and mounting tolerances.

Digital imaging is a large category of an assortment of printing technologies, including electrophotography, electrostatic, thermal transfer, and inkjet printing. Their printed images are difficult to identify by sight and nearly impossible to tell apart in relation to their heat sensitivities. Digital printing ranges from desktop printers to copying machines, where a desktop printer requires a stream of digital data, while a copier requires an original hard copy document.

DIGITAL TECHNOLOGIES
ELECTROPHOTOGRAPHY

This group of printers includes black-and-white toner, and 4-color toner copiers, both used to reprint existing original documents. Electrophotography involves the use of electrical, chemical, or photographic techniques to copy previously printed documents. There are wet methods of electrostatic photocopying that use liquid ink and dry methods that use dry granular ink called toner. Electrophotography is based on a dry copying process introduced in 1950 called *xerography*, from the Greek *to write dry*. Xerography is an indirect printing process in which an electrically charged rotating drum receives an illuminated image that has been converted into a dot pattern. It is contacted by dry powder toner which is attracted to the charged drum, which is rolled onto clean paper and is fixed by heat and pressure rollers, also called fuser rollers, the same way a laser printer does.

Color electrophotography includes all basic 4-color copiers, as found in office supply stores. The 4-color terminology refers to the four separate toners used to form the color images: cyan, magenta, yellow and black. To create color images, the entire process described above for black toner must be repeated four times, each time corresponding to each individual toner. The illuminated image is internally color separated and the image drum is activated and rotated once for each color separation. After the drum has rotated four times, the paper is fed and the image is transferred onto the paper, which is fixed by heat and pressure.

This process was originally launched by Canon, Inc. in the early 1990s, and remains the primary system for color copiers and printers in offices. In its infancy, this process was significantly prone to light fading and color damage by the application of heat and laminates. Today tests have proven these copies to be nearly lightfast.

ELECTROSTATIC PRINTING
Electrostatic printing uses printers that use pigment base toner on dielectric paper and laser printers, and is generally not used for fine art. It is a copying process that uses static electricity (electrostatic) or the attractive force of electric charges to transfer the image to a charged plate or drum. A laser gives a negative electric charge to a cylinder in the specified pattern corresponding to the image to be printed. Positively charged toner is attracted to negatively charged areas on the drum. Paper is pressed against the drum, receiving the toner and is run through heat fuser rollers to set the image.

Electrostatic graphics are defined by the fact that images must be printed on special dielectric media, usually paper. The machine applies the electrically charged pattern to the surface of the dielectric paper. Once printed the image may be transferred to another substrate such as vinyl, then laminated for commercial wide format use. This transfer process falls into three categories: image transfer (for outdoor signage), dry transfer (short to mid-term use banners), and wet transfer (numerous commercial applications). Color electrostatic plotters are considered the top choice for fast, accurate, high-quality color images. These are large format printers for commercial applications.

Although electrostatic photocopying does not use thermal papers, however it does use a heat-set ink process. It is best to select cold mounting techniques of wet, spray, or P-S methods for mounting electrostatic images. Any cold mounting procedure may be used, but moisture control should be implemented. Though relatively lightfast, printed images for commercial use should be cold laminated or surface coated to protect them from moisture, and air pollutants.

THERMAL TRANSFER
This printing technology includes 4-color printers using dyes and pigments on a wax ribbon and a wax-like paper. In thermal printers, a head is in direct contact with the uncoated side of a wax ribbon with the inked side of the ribbon in contact with the printing surface. Ink is heated causing it to melt and adhere to the print surface. Wax ribbons in each of the four process colors (CMYK) pass over the receiving media while a thermal printhead lays down minute dots of wax in precise color locations.

Thermal printing produces uniform dots and color densities. Spot-color ribbons have good opacity and consistent color characteristics. The resulting images may be 40" wide, UV and moisture resistant. The advantage of thermal transfer technology is its tolerance for outdoor use without lamination or transfer to another substrate. Since the image is created with the use of heat, the resulting prints can be heat-sensitive. However, this is not always the case, considering the hottest temperatures used in the process are significantly higher than those used with traditional mounting equipment and adhesives.

DYE SUBLIMATION

Thermal dye sublimation technology of the 90s is a standard for photo realistic printers as a common thermal transfer process. Created by Dupont, dye sublimation is a two step process by which dye images are printed to a carrier as a mirror image, then printed to another (typically poly-based textile) through heat and pressure. Dyes are gasified (vaporized or sublimed) and absorbed by the receiving medium. Dye sublimation printers have the ability to create continuous tone images of varying dot size, producing image crispness and near photographic results.

Inkjet vs. Thermal Transfer

Commercial inkjet printers differ from thermal transfer printers in that inkjet sprays droplets of ink through nozzles onto the print surface. Thermal transfer printers use a head containing resistive elements and ink-coated ribbons, in direct contact with the uncoated side of the ribbon while the ink-coated side contacts the printing surface. The ink is heated, melts and adheres and transfers to the printing surface.

In thermal transfer, the amount of heat determines the amount of wax applied and the density formed on the substrate. Changing the degree of heat results in changing color density, creating the range of tones of the image. In inkjet, there is only one ink density so tone is achieved only through the number of dots per inch.

INKJET PRINTING

Inkjet printers use liquid inks that are sprayed as a dot pattern onto assorted substrates to produce an image. Inkjet imaging is by far the most common digital device. Printers are relatively inexpensive, reliable, and easy to use. They provide superior color intensity and consistence, but still struggle with issues of image stability, related to light fading, gas fading, and lack of moisture resistance. Inkjet printers use cartridge inks containing either dyes or pigments. They are found in small format as home/office printers; large format for commercial production; and fine art applications.

Commercial inkjet printers use a specially designed substrate rather than the plain paper of home/office printers. Inkjet equipment is less expensive than electrostatic equipment, but are slower to print and the materials are higher cost. Since there are so many variations and types of inkjet, their individual heat sensitivity varies depending upon substrate, inks and printer combinations. Average dry mounting heat of 185°F should be avoided, but low temperature HA mount boards of 150°F may be safe.

There are two basic categories of inkjet printer, drop-on-demand (DOD) and continuous flow. DOD technology deposits ink as needed, at the appropriate location as the print head moves back and forth across the paper surface. There are three types of DOD printers: thermal also known as bubblejet, phase change, and piezoelectric. Continuous flow sprays ink continuously towards the printing surface with unwanted ink deflected from it .

PHASE CHANGE
Phase change is a type of inkjet which uses a solid wax-base CMYK color ink stick or puck. The wax ink is heated then projected and applied through the nozzle onto the paper. This type of printing is more commercial and will print on many types of substrate.

PIEZOELECTRIC
Piezoelectric printing, also known as *micropiezo*, uses ink droplets that are squeezed through a nozzle when voltage is applied to a crystal. The crystal pushes on a sealed membrane that pushes the drop through the nozzle onto the substrate. A separate crystal is used for each color (most require six) and a drop is pushed out for application with every voltage signal.

Piezoelectric printers are used predominantly in fine art, large format printing and are manufactured by leading names including Epson, Xerox, Tektronics, Roland and ColorSpan. They print with water or solvent-base inks, containing dye and pigment.

THERMAL (BUBBLEJET)
Thermal is the most common desktop technology. In the thermal print head the ink is drawn into a chamber where it is heated, pressurized, and jetted through a nozzle onto the paper, canvas or other media. Thermal DOD printers are generally not used in fine art printing of giclees'. Contrary to their use by some artists for in-house small-run limited editions, images produced from this type of printer should not really be considered as a fine art giclee or controlled limited edition. This is not to say that computer generated originals are not fine art. In that case the printer and computer are tools of the artist.

CONTINUOUS FLOW
Noted for near photographic duplication, continuous flow inkjet printers such as the Iris printer have revolutionized the fine art industry with such a tight dot pattern it appears nearly to be a continuous flood of ink. Iris systems developed a 3-picoliter droplet, which makes 300dpi appear to be 2000dpi. The technology is of such notable quality that Iris prints are on exhibit at the Museum of Modern Art, NY.

Other companies have now followed in the footsteps of the high tech age of continuous flow printers first established by IRIS and with wide format printers and lightfast pigmented inks, the quality of digitals is soaring.

> ***TIP:*** *MOUNTING DIGITAL IMAGES*
> Some digital photos may be heat tolerant as with dye sublimation prints, while others may be heat-sensitive inkjet prints. Fingerprints left on a digital photo during handling can turn red when heat is applied. Cold mounting with pressure-sensitives is suggested to prevent damage.

ILFOCHROME CLASSICS

Cibachromes are no more like a standard RC (resin coated) photo than a Polaroid. Ilford Color Products set the standards for production of color prints and transparencies through the development of its Cibachrome process. In 1991, Ilford changed the name from *Cibachrome* to *Ilfochrome Classic*.

Glossy Ilfochrome Classics are based on a unique silver dye bleach system using azo dyes which are incorporated during manufacturing, different from RC images which form dyes from color couplers during developing. Cibachromes therefore offer excellent image stability, crispness, and the most lightfast color photo process currently available. With proper developing, handling, and storage, Ilford guarantees the process against color shift for 200 years. The Deluxe Glossy surface materials have an opaque white polyester base 7 mils thick coated on the back with a matte gelatin layer. The 100% polyester base makes the photo print more dimensionally stable while the gelatin backing equalizes the surface tensions making them lie flat.

The challenge is to retain the smooth surface of the Cibachrome without the interference of orange peel. This is when the surface of a photo reflects the visual texture of the underlying substrate, appearing lumpy. Selection of the proper substrate helps minimize the textural appearance. Aluminum and plastics such as Lexan, make excellent surfaces because of their smooth nature and dimensional stability (will not expand or contract with humidity).

Other possibilities include mat boards, paper-based mount boards, and foam center boards, though they are all prone to orange peel. Rigid surfaced boards, including Gatorfoam, Ryno-board, hardboard, and MDF, may be smooth but some might outgas. Regardless of the selected substrate, remember the total amount of pressure applied during mounting impacts the degree of visible orange peel.

Cibachromes are extremely susceptible to damage from mishandling and special care must be used when handling them, as with all photographs. Wearing white photo gloves both at the design table and during framing will prevent oils and fingerprints. Moisture from one's own breath when blowing away dust can leave permanent water spots, and wiping the surface of a Cibachrome with a rag may cause permanent scratches. Treat them with dignity like the valuable artwork they are.

Besides Ilfochrome Classics, Ilford makes an extensive line of additionally Cibachrome print materials from transparency display films to RC emulsions. The same Cibachrome emulsion as above is available as a Semi-matte or Pearl Surface finish on a resin-coated paper core base, making it an RC photograph. Though a beautiful image, it does not display the same high gloss, long term color permanence as the polyester-based materials described above. These RC Cibachromes may be mounted, laminated or canvas transferred as any RC photo.

ILFOCHROME CLASSIC: *STATIC MOUNTING*

There are better solutions to physically mounting a Cibachrome, which maintains the initial dignity of the pure Ilfochrome Classic high gloss appearance. Using archival corner pockets; edge strips; T-, pendant, or flange hinges are the good solutions. Hinging remains the most reversible and noninvasive method for mounting, and hinging in connection with use of static electricity can maintain the image dignity while effectively holding the photo in place.

Cibachromes are flat but floppy. The larger the print the more there is a tendency to buckle from their own weight, potentially slipping out of upper corner pockets and slouching in the center of the mat window. Larger images (16x20" and up) also suck to the surface glazing due to static electricity of the polyester.

MATERIALS
      11x14" or larger Ilfochrome Classic (not RC Cibachrome)
      1/8" Plexiglas or acrylic sheeting with paper covering
      Acid-free hinging tape
      Mat(s) to complete mounting

1. Size all materials, including Plexiglas.
2. With gloved hands, align photo on paper covered acrylic, mark with pencil.
3. Cut through paper 1/64" larger than pencil line.
4. Peel paper cover from acrylic.
5. Flange hinge photo into cut out area,
   settling into small-scale sink mat made from protective acrylic paper.
6. Cover with mat 1/4" over loose photo to hold down.

Small Ilfochromes will hold perfectly with the static and window mat ATG taped to it, larger prints are slightly more secure with a full top hinge. It is recommended to apply the acid-free pressure-sensitive flange hinge across the entire top of the photo, not just in typical pendant hinge locations. Flange hinging is applying a folded V-hinge across the entire top of a hinged item. Nothing will soak into the polyester and there is no moisture to buckle it.

Removing the photo from the mounting only requires the mat to be removed, the flange will come off with water, and lifting the photo breaks the static.

NOTE: This process will not work with a paper based RC Cibachrome photograph. Static electricity will only hold plastic to plastic.

STATIC MOUNTING STEP-BY-STEP

Size acrylic sheet and mark at desired centered position for placement of Cibachrome.

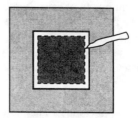

Hold photo in proper centered position, mark with pencil, cut 1/32" larger than pencil line.

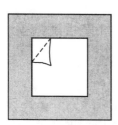

Peel inner paper from plastic to expose acrylic and create natural static. Leave outer paper intact.

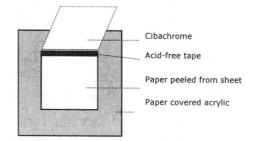

Align photo at top edge with back edge visible. Apply acid-free P-S tape as a flange or v-hinge along top. Fold photo down into exposed window and it will be supported as a mini sink mat, see below.

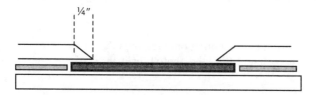

The static will keep the polyester Cibachrome sucked back against the acrylic sheet to prevent sagging forward into the mat window. The overlapped mat window will keep the photo edges held down maintaining the static. Be sure to overlap mat window ¼" or more.

9

Dry Mounting

As discussed in Chapter 8, a cold vacuum *frame* is a vacuum system using no heat. The phrase *vacuum press* generally denotes the use of heat when discussing mounting or laminating, and any mechanical, vacuum or laminating equipment. Although spray and wet glues may be used in these presses, dry mount heat-activated adhesives, films and laminates should be used with heat systems and generally will create a more lasting bond. The ability to use heat within the combo vacuum press not only ensures all aspects of laminating and creativity at your fingertips, but the very essence of it being a vacuum system also provides the benefits of consistent mounting quality from job to job and operator to operator.

BASIC PRINCIPLES

Dry mounting is the application of artwork or fabric to a substrate using heat-activated adhesives in a pressure controlled piece of equipment. Dry mounting is the most time effective, clean, foolproof method for mounting and by far the most permanent, aside from conservation techniques.

TIME
The time it takes to dry mount a project will vary depending upon the adhesive, mounting size, thickness, selected temperature, and item being mounted.

Dwell time is the time the artwork remains in the press to adequately heat all materials, activate, and create the bond. Average dry mount time is 4 minutes for a vacuum press and 1-2 minutes for a mechanical press.

TEMPERATURE
All adhesives have manufacturer suggested temperatures for use in order to achieve best results. There is not a standard, ideal temperature to be used in every heat-mounted situation. Average temperatures run 180-190°F.

There are two basic types of heat producing dry mount presses, mechanical and hot vacuum. One is not better than the other, only different, and selection of the right press to suit your needs varies in every case.

The most obvious difference between basic equipment types involves the elements of time, temperature, pressure and moisture (TTPM).

Mechanical presses were designed to control only two of these elements while the hot vacuum press assists in controlling all of them. There are pros and cons to both systems so select a press based on projected workload today, as well as five years from now.

TIPS AND TECHNIQUES
Basic dry mounting whether in a mechanical or hot vacuum press is simple and consistent. Set the time and temperature based upon manufacturer specifications for selected adhesives and make certain the press is warmed to the proper temperature prior to mounting. The standard correct mounting package from top to bottom includes:
>Release paper of choice
>>Single or double-sided release paper
>>Clear Mylar/polyester release film
>>Release board
>Artwork
>Adhesive
>Substrate
>Release paper (not release board)

PRESSURE
Pressure is the force that squeezes the air from between the substrate, adhesive, and artwork being mounted; holding them in place while the bond is created. A mechanical press must be manually set in order to apply the appropriate pressure for the thickness of the substrate. A vacuum press conforms to various substrates without manual adjustment.

MOISTURE
When existing moisture is heated in a press, the vapors that develop may become trapped between the layers being mounted. Predrying materials is required when mounting in a mechanical or hardbed press. In a vacuum press, moisture is pulled out automatically.

WEIGHTING
Though the selected adhesive may have bonded in the press as it reached temperature, cooling under a weight expedites cooling, reflattens the substrate and will reinforce good mounting practices. Removable tissues bond as they cool, making a weighted cooling station mandatory. A 1/4" thick tempered plate glass slightly larger than the press platen makes a perfect weight. It is cool, heavy, and may be used as a cutting surface.

Allow the front edge of the glass to stick beyond the edge of the work table 1". Screw a 1x2" board at the back edge as a stop. This allows for easy lifting.

TACKING BASICS

Tacking is a process of temporarily positioning artwork onto the substrate so the items being mounted do not shift or move out of alignment as they are moved into the press or as the vacuum is drawn. Using a heated tacking iron is the standard suggested method for spot activating the heat-sensitive adhesive holding art and adhesive to board.

SURFACE TACKING
This is affixing mounting layers from the face of the project through to the substrate. Some mounting requires this technique, such as multiple biting, in order to help prevent buckling of the layers.

1. Position the art with a sheet of adhesive beneath it onto the chosen substrate.

2. Cover the spot to be tacked with a small piece of release paper to protect the surface of the artwork.

3. Using a small circular motion the size of a nickel, tack the print for about 5 seconds with a tacking iron set on medium heat. This will activate the adhesive only at that point to hold all the loose pieces from shifting as it is moved to the press for actual mounting.

Tacking must be done in one spot only, along the end or side of the item to be mounted. Do not tack along an entire edge, at all four corners, in opposite corners or in the center. This could inhibit the paper during expansion/contraction when adjusting to temperature and humidity changes while being mounted and create permanent wrinkles or creases in the mounting.

DO tack

DO NOT tack

Z-METHOD TACKING

An alternative method to surface tacking through all the layers at one time is called *Z-method*. Photo emulsions, copier art and some prints are effected by excess heat and could produce shiny spots during the tacking process if done from the face of the art. Z-method tacking is designed for delicate items so the iron never comes in contact with the surface of the print or photo. If the tacking iron has been designated for framing and has a rheostat, never set it higher than 2/3. Framing irons heat to high 200°F. There are various irons used in the hobby industry for model airplane coatings that reach into the high 300°F temperatures that will damage art. By keeping temperatures low, there is less potential for problems. Always use a slip of release paper between the iron and tacking surface.

1. Tack adhesive to the back edge/end of the photo through a piece of release paper.

2. Align the photo/print to the board as desired.

3. Hold the art in place with the light pressure of the heel of your glove covered hand, and lift the free edge of the photo to expose the adhesive beneath and the substrate.

4. Tack the adhesive to the substrate on the end opposite the first tack point.

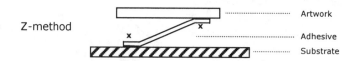

MULTIPLE BITE TACKING

Multiple bites is the process of mounting oversized art, photos, or fabrics in a mechanical press by making a series of smaller mountings or *bites*. The art must be surface tacked in only one place rather than the recommended Z-method. This will prevent any possibility of buckled layers during bites. It is essential to use a permanent adhesive that bonds as it reaches temperature, to prevent adhesive reactivation with successive bites.

The portion of the poster to enter the press first is where the tack should be made. If a piece is meant to be pressed twice (two bites), tack on the <u>end</u> to be inserted first. Always turn the mount board completely around to mount the other side, even if the substrate is narrow enough to be pushed straight through. The board may not always fit between the back braces of the press and damage may be done to a soft foam board.

If the poster needs four bites, tack on the **side of the quarter** to enter the press first. With a four step mounting it does not matter whether you move adjacent or across from the initial mounting (clockwise or counterclockwise), but always be systematic and complete the entire project once begun.

In a longer poster, still narrow enough to fit within the confines of twice the platen width, tack in the **center of one of the long sides**. This then will be fed into the press first, then directly across from the initial mounting. Then move either right or left from the center to the far ends, making it much easier to ensure proper placement on the mount board as well as pressing the air from the center to the outer perimeters of the poster to prevent wrinkles or creases.

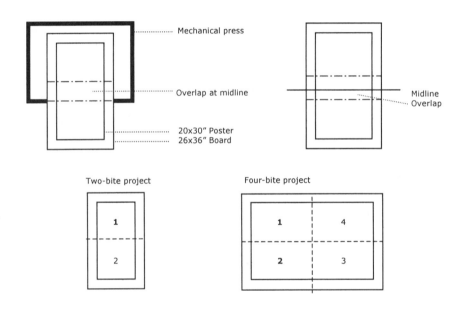

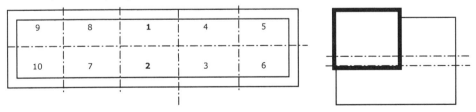

Overlap is allowed a little less than twice as wide as the press width.

MECHANICAL PRESSES

Time and temperature are variables that will change with individual projects and selected adhesives. The pressure and moisture elements must be manually controlled when using a mechanical press.

In a mechanical softbed press, the lever arm must be properly adjusted for the thickness of the substrate or there is a risk of a poor, unpredictable, inconsistent mountings and possible dents in the board. Dry mounting is a predictable and reliable form of mounting as long as the elements of time, temperature, PRESSURE and MOISTURE are properly controlled.

CHECKING THE PRESSURE

1. Cut a 20" square piece of foam board, and score diagonally corner-to-corner only cutting half way through the thickness of the foam. Fold the square in half to create a perfect 45-degree pattern. By using this large a piece of foam it is easy to visually match the angle of the pattern against the press handle (lever arm) plus it remains self-supporting.

2. Place a piece of 3/16" foam board in the press with all other portions of the mounting package including the release materials. Close the handle (or lever arm) to the resting but not locked position and match it to the pattern above.

3. The lever arm should match the 45-degrees of the pattern. If it is too high the press will clamp too tightly, if too low there will not be adequate pressure to ensure a proper bond.

The standard pressure setting of a mechanical press is for 3/16" foam boards, but set it for the thickest board to be mounted.

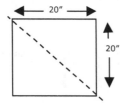
Score a 20x20" rectangle diagonally

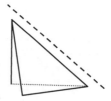
Fold into 45-degree self-standing angle

ADJUSTING FOR PRESSURE

The *locking nuts* are the doughnut shaped outer rings that loosen when turned counterclockwise to adjust the taller, inner *pressure adjusting screws*. If the locking nut is too tight to loosen, insert a 3/16" foam board for space, then lock the arm down to loosen the pressure of the locking nuts on the press, and unscrew them until they are out of the way or off.

1. Loosen the locking nuts.

   .... Adjusting screw
   .... Locking nut

2. Place a 3/16" foam and release board to loosen the inner pressure and lock press closed, if necessary.

3. Twist the tall adjusting screws to raise or lower the lever arm to accommodate the inner mounting materials, checking for 45-degree angle.

4. Lock the press lever arm down and replace the locking nuts finger tight.

Twisting the adjusting screws evenly to the right or in a clockwise position will drop the lever arm down, thus lightening or decreasing the press pressure. Twisting the nuts to the left or counterclockwise will raise the arm up and increase the press pressure. Turn both screws at the same time when adjusting. Check with owner's manual.

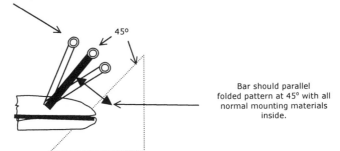

Bar too high creates too much pressure, too low not enough pressure.

Bar should parallel folded pattern at 45° with all normal mounting materials inside.

Mechanical softbed press requires proper manual adjustment for correct pressure.

The three basic bar positions:
- All the way to the back of the press is the *open* position.
- Resting at 45° to the table is *closed*.
- Clamped down tight and horizontal to the table is *locked*.

SHIMMING FOR PRESSURE VARIATIONS

Readjustment of the press is required each time the substrate changes or the pressure will also change. As an alternative to physically readjusting the locking screws, shims may be cut to accommodate the variation in substrate thickness and placed beneath the Masonite board to make up the difference in thickness. The shims are placed beneath the Masonite, not within the platen/sponge pad mounting area.

1. Place a strapping tape tongue on the front edge of the lower Masonite pressure board for easy accessibility.

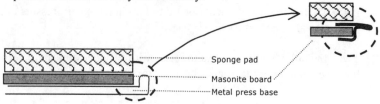

2. Cut three 4-ply equivalent scrap mat or mount boards about 1" smaller than the size of the sponge pad or Masonite.

3. Slide a shim beneath the Masonite whenever a thinner mounting is required. Remember the press was adjusted for 3/16" foam and 3 shims = 3/16" foam.

4. For quick adjustments:

0 shims for	3/16" foam substrate
1 shim added for	1/8" foam
2 shims added for	4-ply mat board
3 shims added for	when no substrate is used

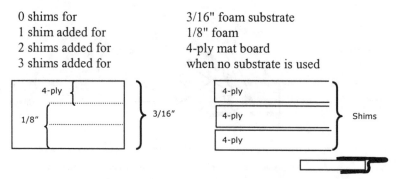

Tab shims with tape for easy removal.

When no solid substrate is used, as in premounting or with a canvas transfer, three shims will equal the entire 3/16" inner press adjustment.

HARDBED PRESSES

Mechanical presses discussed in this book are known as *softbed* presses outside the United States. *Hardbed presses* are the European version of a mechanical press. They come in two average sizes of 16x16" and 24x24" have a hard metal base, ¼" metal platen surface plate, and wheel on top to tighten for pressure. It uses a thin aluminum sheet on top and a hardboard carrier for mounting. They are thermostatically controlled with temperature readouts and operate on 220V.

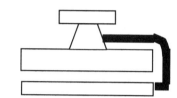

This skeletal line drawing of a hardbed press shows the top wheel that controls the pressure.

They come both as manual wheel and hydraulic. The hydraulic presses are capable of much higher pressures than a vacuum press.

PREDRYING FOR MOISTURE CONTROL

Predrying materials is a very simple process and a mandatory practice when mounting any project using a mechanical press. By removing the moisture from within the board prior to mounting, the adhesive is allowed to fuse to the substrate as a moistened stamp to a dry envelope.

1. Fold a clean piece of absorbent Kraft-type paper in half as an envelope.

2. Place the artwork or material to be mounted into the envelope then into a closed, but not locked, press for 15-30 seconds. The substrate and art must be predried separately.

3. Do not predry using a release paper envelope instead of Kraft paper, the moisture will be trapped and turned to steam.

> ***TIP: CONTROLLING STEAM***
> As residual moisture in any of the mounting materials reaches 212°F, it is turned to steam. That steam expands with greater pressure than the original moisture making bubbles a bigger problem. Always remember to predry.

HOT VACUUM PRESSES

A heat vacuum press applies uniform heat and pressure as the art is mounted. A vacuum press is limited to mounting items, which fit totally inside its frame disallowing multiple bite options. But they are available as much larger units to comfortably accommodate full sized 32x40", 40x60", and 4x8' boards in one single step.

Dry mounting denotes the use of heat when discussing the subjects of mounting and laminating, because dry mount tissues, films and laminates require heat for bonding. On occasion, spray and wet glues may be used in hot vacuum presses if they are used without the heating element turned on as in a cold vacuum frame. This is a nice option when pressure is required without heat for more heat-sensitive projects.

TECHNIQUE
The elements of time and temperature must still be monitored by the press operator based upon selected adhesives, substrate, and overall size of the mounting project, but the elements of pressure and moisture will be controlled automatically by the heat vacuum system.

Routine maintenance of release papers, overlay foams, and platens is necessary for either mechanical or vacuum systems. Other general maintenance considerations include the pumps of various vacuum presses, which suggest running units up to 5 minutes prior to shut down each day to clean the vacuum system of accumulated moisture. Regularly check the pump for moisture buildup, clean filters, and oil the pumps as needed. All manufacturers' manuals should be closely read before initial operation.

In opposition to the mandatory procedure of predrying when using a mechanical press, the vacuum principle of drawing the air from within the press precludes this step when using a hot or cold vacuum press of any kind. The physical drawing of the vacuum pulls the moisture from the materials automatically, while the rubber diaphragm or bladder naturally conforms to the thickness of each individual substrate adjusting for pressure.

PRESS COMPARISONS
The actual pressure of a mechanical softbed press averages 2-4psi whereas a hot/cold combination vacuum press is approximately 12-14psi (24-28"Hg). This lower poundage in no way inhibits the ability of a mechanical press to produce as good a mounting as a vacuum press. Hardbed presses average 4-5psi with average wheel tightening, though additional rotations continue to increase to 2 times vacuum poundage. Hydraulic presses claim 500psi.

See Appendix for Pros and Cons of Selecting a Press.

REVIEW OF HINTS AND REMINDERS

- Never store two-sided release paper rolls next to roll tissue adhesive. They look very similar and the adhesive could be mistaken for release paper.

- Release papers cannot be used for predrying because they cannot absorb moisture. Excessive moisture in a release paper envelope can badly cockle and warp from the presence of moisture, which in turn may effect the silicone surface.

- Release materials have a lifetime of around fifty working hours. They will only produce smooth, clean results if they are not attempting to transfer their own wrinkles of age or damage.

- Using a release envelope is an economic, sensible and systematic use of materials (rather than full-size press sheets) when mounting smaller items.

- Single-sided release paper is more rigid, not as slippery and works extremely well as a wrap for the sponge pad of a mechanical press. Cut the paper width of the pad, fold the excess release paper around the ends of the pad, and tuck between the pad and the Masonite.

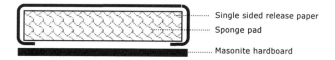

- Brown Kraft paper is not a reasonable substitute for release paper, although it is necessary for predrying when using a mechanical press. Kraft paper could glue itself to the art or platen if any glue residue has been left on it. Also, heated inks could stick or transfer to the paper.

- Do not use release boards as the bottom of a mounting package. The substrate must compress into the mechanical press pad or allow diaphragm to contour around it. If a board is used both top and bottom, excess pressure can occur at the outer edges as the sponge or diaphragm attempts to conform to the shape of the desired inner project, thus bowing the board and possibly creating uneven mounting pressure in the project center.

80 / *The Mounting And Laminating Handbook*

- Transference of adhesive residue and wrinkles from previously mounted projects may readily appear if overworked release papers are used. A board will begin to brown slightly with extended use, and its lifetime is somewhere around fifty hours of production time.

- Using two thin pieces of clear polyester release film or double-sided release paper (in a vacuum press) can create a suction sealed envelope if the sheets are much larger than a small project. Trapped air will appear to be a lack of pressure when in fact it's the suction that won't allow air to escape.

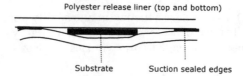

- A single oversized clay coated foam board can also create a sealed suction to the platen or to thin release materials if too much is left unmounted surrounding the project. This will trap air and create uneven mount pressure. (see above diagram)

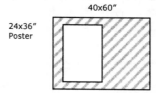

- As moisture reaches a temperature of 212°F it turns to steam that in turn expands creating greater pressure in any contained area. Steam bubbles trapped beneath nonporous art will never go away.

- Countermounting the back of the substrate will reduce bowing of the board. The same size and weight of adhesive and paper must be used to equalize the surface pressure from both sides.

➤ *TIP: COMPUTER ART*
The same as any digitally produced image, consideration must be made for value, edition, and technical process to determine mounting. Even if they are heat tolerant they may be considered originals requiring preservation mounting.

10

Dry Mounting Applications

BASIC TECHNIQUES

The applications presented in this chapter are designed to teach basic dry mounting techniques, rather than specific projects, so that any project may be tackled. Learning how to become more time and material efficient by understanding terminology and procedure will save money and build confidence. These are the standardized and accepted methods of application in the framing industry and adapt to any heat mounting system whether mechanical softbed, hardbed or vacuum press.

Release papers should always be used and cut to appropriate sizes depending upon the press and application. They may be folded into an envelope folder or used as individual sheets, but are used to protect the heat platen in the top of the press and the sponge pad or rubber diaphragm on the bottom. A release board should not be used on the bottom unless specifically noted.

STANDARD MOUNTING PACKAGE

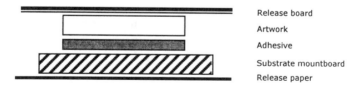

Release board
Artwork
Adhesive
Substrate mountboard
Release paper

Making an envelope of folded release paper allows for easy handling of small projects and those with loose items.

Premounting to tissues is best achieved in a folded release envelope.

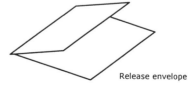

Release envelope

ADHESIVE TRIMMED TO SIZE

Tissue adhesives and substrates may expand slightly in the press when exposed to heat. Therefore, if adhesives are trimmed to the exact size of the art, they will often show beyond the art edge once mounted. If a mat is to be placed over the mounting this is not a problem, but the shiny adhesive would be unacceptable for a plain or center float mount. Trimming the adhesive slightly smaller, (approximately 1/8") will eliminate this problem. Cutting adhesive the exact size of the item to be mounted, then centered, tacked, and mounted can be a time consuming process. It is also not the most efficient method of preventing adhesive crawl.

MATERIALS
Breathable poster print
Substrate of choice
Permanent, porous, tissue adhesive
Tacking iron
Release paper or polyester

1. Preheat press to 185°F.
2. Predry poster and substrate separately, 15 seconds each in a closed but not locked press in a Kraft paper envelope. This step is not necessary in a vacuum press due to draw of the vacuum removing material moisture.
3. Trim sheet adhesive slightly smaller (1/8" less) than print.
4. Position print and adhesive on substrate and tack at one point along the side or edge.
5. Assemble mounting package top to bottom:
   Release paper
   Poster
   Tissue adhesive
   Substrate
   Release paper
6. Place package in press allotted time, 1-3 minutes mechanical, 3-5 vacuum press.
7. Remove from press and cool under weight.

OVERSIZED ADHESIVE FOR MATTED ART *(technique next page)*

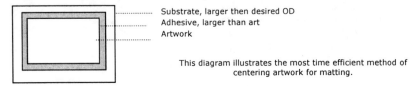

Substrate, larger then desired OD
Adhesive, larger than art
Artwork

This diagram illustrates the most time efficient method of centering artwork for matting.

OVERSIZED ADHESIVE FOR MATTED ART

The most time efficient way to dry mount artwork slated for matting is by working with slightly oversized tissue adhesive. Cutting adhesive larger than the artwork, but smaller than the substrate, saves measuring and alignment time prior to mounting. In this case, also use a slightly oversized substrate for the project (1-2" around), which saves time when aligning the completed window mat.

MATERIALS

- Porous poster print to be matted
- Substrate of choice
- Permanent, porous, tissue adhesive
- Tacking iron
- Release paper

Cut all materials slightly larger than desired size, and cut window mat opening.

1. Preheat press to 185°F.
2. Cut the substrate 1-2" larger than outside dimensions of the mat, and the adhesive larger than the print but smaller than the substrate.
3. Predry poster and substrate if using a mechanical press.
4. Tack the edge of the adhesive and print to the visual center of substrate. The oversized substrate allows for only a visual centering rather than taking time to measure for exact center placement.
5. Assemble mounting package top to bottom:
   - Release material
   - Art
   - Adhesive
   - Substrate
   - Release material
6. Place in press 3-5 minutes, time designated by size of project and type of press.
7. Cool under weight.
8. Cut window mat, line with ATG tape, then place into position on top of mounted poster.
9. Trim excess substrate from around outer edge of mat.

OPTION: After mounting, position mat, line with pencil at outer perimeter of mat, set aside, trim away substrate, and assemble for framing.

PREMOUNTED ADHESIVE

Since tissue adhesives may slightly expand in a heat press during mounting, tissues cut exactly to size will often show a thin adhesive line around the outer edges after mounting. If, however, the items are premounted onto an oversized adhesive sheet and then cut to size or shape, no adhesive will expand beyond the outer edges during the final mounting application.

MATERIALS
- Artwork or photo
- No substrate
- Permanent, porous, tissue adhesive
- Tacking iron
- Scrap mat board slightly larger than adhesive
- Release paper

1. Predry items as necessary for mechanical press preparation.
2. Cut tissue adhesive larger than the art.
3. Tack the edge of the art randomly to the adhesive.
4. Place in a folded release paper envelope, as the adhesive will hold-on to the paper while hot, making it difficult to remove from the press.
5. Mount at 185°F for 1 minute mechanical press, 2-3 minutes vacuum press.
6. Cool under weight to keep flat.
7. Trim to exact desired size or shape of art.
8. The artwork is now backed with heat-activated adhesive ready for completion of project. Any tissue expansion from heat will have occurred during premounting. So once trimmed clean the adhesive will never show around the mounted art. Tack to substrate and place into press second time for final mounting.

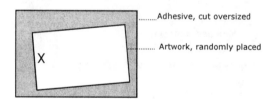

> **TIP: VACUUM PRESS BOARD STIFFENER**
> During mounting, a mat board stiffener beneath the mounting envelope takes the place of missing substrate in a vacuum press, protecting the project from diaphragm wrinkles.

MONTAGE OR MULTIPLE MOUNTINGS

When a project requires numerous small items to be mounted to a colored backing, the best dry mounting solution would be premounting. Then the adhesive backed pieces may be arranged, stacked, tacked, and mounted to create the finished project. Unusual shapes are may be cut from the adhesive sheet with knife and self-healing mat with no fear of adhesive showing around the edges.

MATERIALS

    Assorted odd shaped items such as valentines, wine labels, photos, or clippings for montage
    Colored substrate
    Permanent, porous, tissue adhesive
    Tacking iron
    Scrap mat board slightly larger than adhesive
    Release paper

1. Heat press to 185-190°F and predry items as necessary for mechanical press.
2. Cut tissue adhesive large enough to fit randomly placed items.
3. With release paper beneath, tack all items to adhesive.
4. Mount in press using release paper envelope, with a mat board stiffener beneath the envelope (not inside) if in a vacuum press.
5. Mount 1-2 minutes, cool, and trim out individual items.
6. Tack all trimmed items onto desired colored substrate.
7. Mount a second time, 185°F, 2-5 minutes, cool under weight.

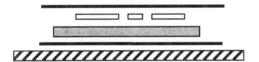

Release paper, **not board**
Artwork
Adhesive
Release paper
Rigid board stiffener (vacuum press only)

Premounting oddly shaped items or a group of the same (like photos) is the best way to affix adhesive to them prior to mounting.

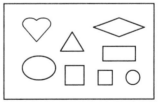

PLAIN OR FLOAT MOUNTING

Plain or *float mounts* are the most basic forms of print presentation and protection. When a print or photograph is premounted with a permanent adhesive, trimmed to remove outer adhesive to size, then mounted centered onto a colored backing board, it is called a floated mount. Photographs are commonly float mounted as 11x14" photo prints onto a 16x20" display board. Borders are generally 2-3" around and though this is not the preferred method for glazing and framing, it is frequently used for photo storage.

FLOAT MOUNTING FOR PHOTO STORAGE

This is a preferred process for photo storage to maximize protection of photo edges. Though the addition of window mats will best protect the photo surface, the minimal thickness of only one 4-ply board cuts down on bulk. Mounted prints must be interleaved to prevent surface print damage during storage.

Plain mounting requires two times in the press:
1. Premounting of the adhesive to art or photo, at 185°F for 1-2 minutes in a mechanical, or 2-3 minutes in a vacuum, press.
2. Then mount the properly trimmed preadhesived art to the selected substrate, 185°F for 2-4 minutes.

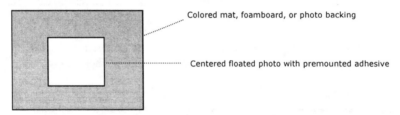

Colored mat, foamboard, or photo backing

Centered floated photo with premounted adhesive

> ***TIP:*** *AIR BUBBLES*
> Bubbles may occur when layering nonporous photos in a montage. Consider a repositionable pressure-sensitive film application that would have more control over trapped air. Overlapping porous papers poses no problem.

> ***TIP:*** *SUCTION SEALING*
> In a vacuum press, release boards placed in the bottom of the press may suck up to the platen surrounding a smaller release envelope. This can restrict the diaphragm from conforming to the substrate allowing for air bubbles. The silicone suction restricts the air from escaping. Never use a release board as a substitute substrate or stiffener.

FLUSH MOUNTING

When a mounted image extends clear to the edge of the substrate and is not matted it is known as *flush mounting*. Originally popular in the 1950s, it has traditionally been used with photographs but may also be done with posters and cards. The process is achieved by either trimming excess white edges from the print prior to mounting or by mounting everything slightly larger, then trimming down after mounting.

The mounted print/board unit may be bevel, reverse bevel, or square cut at the exact photo edge to create increased dimension. If the mounted photo is to be floated within a shadow box design frame, the edges will remain undamaged. If mounting photos for storage, flush mounting defeats the purpose of protecting edges from damage by trimming the board to the image edge, allowing it vulnerable to dents and bending.

Preheat press to 185°F. Cut to size, predry and prepare materials as necessary for mounting. Size substrate about 1" larger than desired image to have extra for trimming. Mount art, photo, or sign with a permanent, breathable, tissue adhesive to the substrate and cool under weight. Mounting time will vary depending upon thickness of substrate and overall size of the board. Standard times will be 1-3 minutes for a mechanical press, 3-5 minutes for a vacuum press.

With a mat cutter, bevel or reverse bevel edges may be easily trimmed to the exact edge of the photo. By selecting either black or white 3/16" foam board as a substrate, this type of bevel trimming creates a narrow contrast trim. Traditional or competition grade photo mount boards will better protect the edges than any foam but are not decorative for bevel cutting.

TYPES OF EDGING

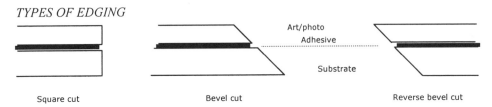

Square cut   Bevel cut   Reverse bevel cut

> **TIP:** *DIGITAL PHOTOGRAPHS*
> Electronically recorded or scanned, manipulated and printed with an inkjet or thermal transfer printer, these images look like traditional RC photos. They may be heat-sensitive suggesting cold mounting methods to prevent damage. Ask for a second image so proper testing may be completed prior to mounting of the art image.

MULTIPLE BITES

A mechanical press, hardbed or softbed, can mount items near twice its width and miles longer then itself using the multiple bite process. This is merely a process of placing a piece into the press more than once to mount sections of the whole in smaller bites. The multiple bite flexibility of a mechanical press is an advantage to the framer who only rarely requires the facilities to mount an occasional 24x36" or oversized poster.

Use a permanent, breathable tissue for multiple bites. These tissues bond under heat within the press and being permanent will not release with subsequent visits to the press for additional bites.

MECHANICAL SOFTBED PRESS
- Adjust press pressure (lessen) to compensate for full depression of sponge.
- Always use a release board to dissipate pressure at the bite point.
- Predry art and substrate.
- Use a permanent, porous mounting tissue.
- Tack poster or photo at center of first bite.
- Mount systematically, from the center out (see Multiple Bites, page 73).
- Overlap each bite as much as possible.
- Cool under weight to maintain maximum flatness.

A mechanical press would be considered a time detriment to a framer who routinely mounts numerous 24x36" posters which would require the above multiple bite method. If routine 32x40" boards are to be used, then perhaps a hot vacuum with a maximum mounting of 32x40" might be the best consideration. Likewise, a framer who expects to regularly mount 40x60" prints may require a larger vacuum unit.

Remember that the same space will be required to accommodate the full 40x60" substrate whether mounting a 40x60" project in a large vacuum or as a multiple bite project in a smaller mechanical press. See multiple bite tacking.

HARDBED PRESS

Manufacturers suggest inserting ½" foam plastic or overlay foam to help dissipate the pressure at the edge of the press. Place the foam between the surface aluminum sheet and the artwork. Also, use a carrier board beneath the substrate.

Top to bottom: Aluminum sheet, ½" foam plastic, artwork, adhesive, substrate, carrier board (hardboard or MDF). The foam plastic and carrier board takes the place of release boards used in softbed presses. Predry as necessary and follow standard basic steps for mounting in bites as in a mechanical press.

TRANSLUCENT MATERIALS
RICE PAPERS AND ONE-SIDED TEXT
This applies to light colored delicate rice and tissue-type papers. By maintaining a low mounting temperature and shorter time, less adhesive will saturate into the already thin artwork to prevent adhesive from bleeding through.

MATERIALS
    Thin light colored artwork, rice paper painting
    White substrate
    Low temperature, removable, porous, tissue
    Tacking iron
    Release paper

1. Set press at lower temperature of 165-175°F.
2. Predry materials for mechanical or hardbed press.
3. Tack adhesive and artwork to substrate.
4. Assemble mounting package.
    Release paper (top)
    Rice paper artwork
    Adhesive
    Substrate
    Release paper (bottom)

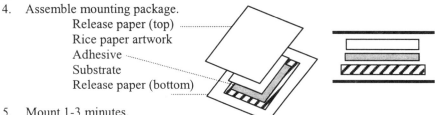

5. Mount 1-3 minutes.
6. Remove from press and cool under weight, remember that removable tissues bond as they cool outside the press.

FACE DOWN METHOD
Adhesives travel up, toward the heat source (platen) of a press, saturating porous materials. In addition to lower temperatures, and shorter press times, turning the project face down when initially mounting will lessen adhesive absorption.

Mounting package from top to bottom:
    Release paper (top)
    Substrate
    Adhesive
    Porous artwork
    Release paper (bottom)

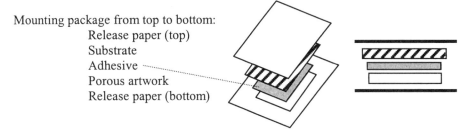

NOTE: Do not mount valuable rice paper originals nor attempt traditional scroll mounting techniques using any mounting method without proper training and years of practice.

NEWSPAPERS AND TWO-SIDED TEXT

Any newsprint, magazine, or flier with two-sided text establishes the potential for ghosting. The dominant color of the verso side will indicate the selection of substrate to camouflage the lettering. The adhesive must be a transparent, nontissue adhesive allowing the dark substrate to blend with the black text on the verso side. This will slightly gray raw newsprint on the facing side. Use of white tissue adhesive negates the benefit of mounting to a dark substrate, allowing the ghosting to appear.

MATERIALS
Newspaper clipping
Black substrate or color of dominant verso lettering
Removable, porous, clear film adhesive
Tacking iron
Release paper

1. Predry materials as needed for mechanical press.
2. Tack adhesive and clipping to substrate.
3. Place in 190°F press for 2-5 minutes
   depending on size and thickness of substrate.
4. Remove from press and cool under weight.

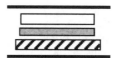

Release paper
Newspaper
Clear adhesive
Substrate
Release paper

The best sales sampler is a 50/50 showing one half mounted to white with ghosting encouraged; one half properly mounted to black to camouflage the verso black text.

➤ **TIP: REPRODUCTIONS**

Due to the high acidity and low quality of newsprint paper, the mounted clipping will yellow and deteriorate rapidly. A good 4-color copy from a Xerox 5775 printed on full color mode will best reproduce the color of the natural newspaper. Since newspapers are copyrighted, a waiver may have to be signed for permission to copy. A photographic replication is also a good alternative to framing the actual original. Then the original can be encapsulated in Mylar and kept in the dark, perhaps even in the same framing package.

SILKS AND SHEER FABRICS

Silks, acetates, chiffon and other sheer fabrics may be successfully dry mounted. Open weave fabrics (chiffon) may allow the melted adhesive to show between the threads once mounted, making the fabric appear shiny. This is not fiber saturation, for dry mount adhesives will not soak through if proper attention is paid to the control of time and temperature. If the fabric is light in color, mount to a light colored substrate. A dark fabric should be mounted to a dark substrate. White adhesive tissue will also effect the substrate by color fading it, clear film adhesive must be used.

MATERIALS

Sheer fabric (chiffon, silk, linen)
Substrate to match background color
Removable, porous, pure film adhesive
Tacking iron
Release materials

1. Predry as needed for mechanical presses.
2. Tack adhesive and fabric to substrate.
3. Place in 190°F press for 1-3 minutes mechanical, 2-4 minutes vacuum.
4. Remove from press and cool under weight.

COLOR TINTING

Color tinting is the deliberate use of the substrate color as part of the surface design. By selecting a red substrate for a thin white document, the resulting mounted document will appear slightly pink. Placing a red substrate behind a sheer black chiffon floral rose print will accent and intensify the red rose, but may alter the black colored background by changing it to a reddish-brown.

A clear pure film adhesive is required for this design application. White tissue will turn a sheer black chiffon middle gray, negating the rich colors altogether. The longer a film adhesive is left in a press for creative applications the greater the saturation. At 190°F for 2-3 minutes the item will simple mount. At 200°F for 15 minutes, it will begin to become transparent due to saturation. After 30 minutes in a press at 200°F, it will saturate a silk threaded rice paper, so only the squiggly fibers and color tint are showcased.

➢ *TIP: SATURATION OF ADHESIVES*

If saturation is extreme (30 minutes) the paper or fabric surface may feel waxy to the touch. If this is a mat design, a spacer should be placed between the glazing and mat.

OPAQUE MATERIALS
FABRIC WRAPPED MATS

Wrapped mats are easily achieved in a heat press. Whether 4, 8, or 12-ply mat, 1/8" or 3/16" foam a heat wrapped mat is quick, clean, and easy. Though any foam board may be used, the toothed surface of acid-free and black (nonbuffered) foam boards, as well as mat boards, hold dry mounted fabrics extremely well. They are the most permanent mounting variation able to survive temperature extremes of humidity and cold without bubbling or lifting over time. Pure film adhesive is required to more easily melt around the bevel and contours of the mat edges.

There are two simple steps required to achieve great wrapped mats in a heat system:
1. Fit the fallout back in place before mounting.
2. Iron the bevel to reactivate the nonmounted adhesive and adhere it to the bevel edge.

MATERIALS
    Fabric of choice (canvas, silk, linen...)
    Mat, Acid-free, or Black foam board
        cut in desired for window mat
    Porous, pure film adhesive
    Tacking iron
    Release materials

TO MOUNT:
1. Size and cut mat or foamboard to desired dimensions, reserve fallout.
2. Size fabric 1" larger then window mat.
3. Assemble mounting package:
   - Release paper (top)
   - Fallout
   - Fabric
   - Adhesive
   - Window mat as substrate
   - Release paper (bottom)
4. Mount at 190-200°F for 1-3 minutes mechanical, 2-4 minutes vacuum press.
5. Remove, cool under weight.

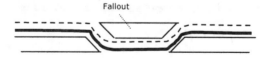
Fallout

TO COMPLETE:
6. Cut inner window opening leaving 1" raw edge, miter the corners, and prepare to reinforce the bevel.
7. Hand iron the bevel edge to mount loose fabric. Foam board is an insulator rather than a conductor of heat; the bevel will not fully mount the fabric due to the fallout insulation.
8. Place mat face down and iron flaps onto back of mat window.
    DO NOT RUSH. Iron slowly for a good bond.

Pay attention to thread pattern in a fabric. Though a soft, stretchy velour will easily wrap around 3/16" foam without visual distortion, a vertically threaded linen will show the curvature dip into the corner on even a 1/8" foam. Select the proper fabric for the desired mat depth.

NOTE: This process is also applicable to cold frames with a wet or spray glue. Do not forget to refit the fallout prior to mounting. Turning the flaps back will require double-sided tape or wet glue to secure.

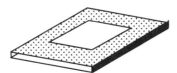

Mount with fallout to establish bevel.
(see diagram previous page)

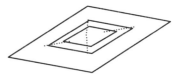

Cut opening in fabric with mat face up. Miter corners. Do not cut too close into the corner, it could expose raw foam or mat board.

Hold the mat with the plane of the bevel parallel to the table and iron the bevel to reinforce and melt adhesive. Foam board is an insulator and will not allow the heat to melt the adhesive when the fallout is in place during initial mounting.

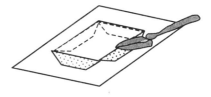

Turn the mat face down and iron the adhesive backed tabs to the back of the mat window. Begin from the center and work into the corners, always pulling towards the outer edge to avoid bevel buckling.

WRINKLED PAPER TECHNIQUE

Decorative papers may also be used for wrapping, but the harder less flexible nature sometimes makes them tough to contour. The wrinkled paper technique softens the fibers of thin, stiff papers allowing them to contour around corners. This process applies well to long fibered Japanese papers as well as thick short fibered Western made papers.

Use the same step-by-step process for wrapping as above. Crush the selected paper into a ball before dry mounting to the window mat, open flat, and then mount with film adhesive. Wrinkling the paper breaks and softens the stiff fibers for easier contour around window openings and embossed mat surface designs.

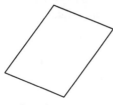 

Smooth art paper   Wrinkled into a ball   Flattened

20# art paper, wrinkled, flattened then mounted to the selected substrate.

> ***TIP:*** *IRONING BEVELS AND TABS*
> A household iron is perfect for bevel ironing and flap turnbacks. It is large and heavy covering greater area at a time. Set just below WOOL will heat up to 190°F to melt the adhesive, when set directly on WOOL, the iron heats beyond 230°F and melts foam board.

> ***TIP***: *THERMOGRAPHIC DARKENING*
> Even when mounted with nonheated pressure-sensitive methods a heat sensitive theater ticket will blacken with time. The inner temperatures of a frame will increase during the day when hung on a room wall exposed to direct sun.

PAPER FIBERS AND PAPER GRAIN

Japanese papers are handmade using long, soft, tough fibers that do not create a natural grain when folded as Western machine-made papers do. Western papers are made of numerous short fibers running dominantly in one direction. A few fibers run at cross pattern to hold the paper together. The *grain* of a paper runs the same direction as the bulk of the fibers, while *cross-grain* is the opposite direction.

Papers and boards may bend, bow or are less resistant to the bend when they are torqued with the grain. This is because of bending fewer fibers in the process. Papers tear easier and straighter when torn with the grain. If a poster is mounted to a board with the grains of both poster and board running the same direction it will encourage the mounting to bow or warp more than if placed at cross-grain.

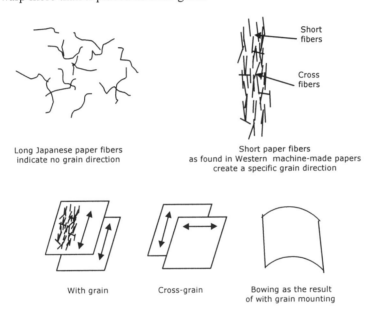

Long Japanese paper fibers
indicate no grain direction

Short paper fibers
as found in Western machine-made papers
create a specific grain direction

With grain   Cross-grain   Bowing as the result
of with grain mounting

Sizing substrates to accommodate the cross-grain patterns of both the poster and the board are an effective way to combat bowed boards when mounted items are to be left unframed. An alternative solution is countermounting. Unfortunately, the greater use of materials, and in turn potential material waste, when ensuring cross-grain may determine whether or the technique is even considered.

DETERMINING GRAIN DIRECTION

Dry Methods = Most resistance to bending is against the grain
             = Smoothest tear direction is with the grain
Wet Method   = Moistened paper scrap curls with the grain

ONE-STEP SHADOW BOX

Shadow boxes are greatly enhanced by fabric application to create a warmer environment for the objects contained. Traditionally wrapped in pieces and then assembled, this variation allows for a quick one-step mounting and assembly.

MATERIALS

- Fabric for box lining
- Foam board substrate (1/8" of 3/16")
- Pure film adhesive
- Tacking iron
- Release paper
- Cork backed ruler or Mat cutter
- Linen tape

1. Size substrate to construct box upon completion according to the following formula. W+H+H-T-T = Width
   L+H+H-T-T = Length
   (W) width or (L) length of box floor
   + (H) height of side
   + (H) height of side
   - (T) thickness of substrate (i.e.: 1/8" or 3/16")
   - (T) thickness of substrate
   = width or length of size of foam base.
2. Score verso side of foam base with vertical cut through backing paper and foam only. Do not bend.
3. Cut fabric 2" larger than foam base around.
4. Predry as needed for mechanical presses.
5. Assemble mounting package:
   - Release paper
   - Fabric
   - Film adhesive
   - Foam box substrate
   - Release paper
6. Tack centered fabric extending beyond all edges of box sides.
7. Mount at 190°F for 1-3 minutes in mechanical, 2-4 minutes in vacuum press, cool under weight.
8. Bend sides to create box shape; the hinges are reinforced by the top paper and fabric.
9. Trim corners out, wrap excess fabric around top raw edge of foam, ATG on box back.
10. Fold up sides and tape into shape with linen tape.

ONE-STEP SHADOW BOX DIAGRAMS

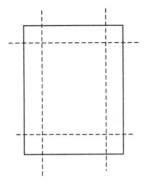

Score the back of the box end to end through the foam, all four sides, down to the top layer of paper.

Do not bend the box sides until after mounting is completed.

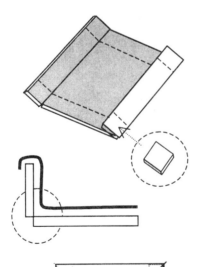

After mounting the fabric or decorative paper, bend sides and ends to create box.

Remove the square pieces from the corners, to leave only front paper and fabric.

Once mounted to the front of the box, the fabric will act as a hinge reinforcer.

There are four ways to finish the corners of a box.

1. Mitered, not cut
2. Miter cut and wrapped
3. Tab cut one side
4. Cut out

JIGSAW PUZZLES

Puzzles may be mounted and laminated using either cold or hot methods to hold them together for framing and protect them from lightfading and other deteriorating elements. Puzzles require two stiff boards to best handle them, a substrate for mounting and stiff secondary board for moving and inverting. Be careful not to lose any of the pieces.

MATERIALS
- Assembled puzzle
- Breathable, pure dry mount film, wet glue or pressure-sensitive adhesive
- Substrate of choice, plus secondary hardboard for turning

DRY MOUNTING
1. Set press temperatures, readjust pressure for added project thickness and predry as needed.
2. Invert puzzle face down on stiff secondary turning board.
3. Size dry film adhesive and substrate.
4. Lay adhesive on puzzle back, substrate and invert.
5. Assemble mounting package:
   - Release paper
   - Puzzle
   - Pure film adhesive
   - Substrate
   - Release paper
6. Mount at 190°F, 4-6 minutes, and cool under weight.
7. Laminating may then be applied at 190-225°F, 5-7 minutes.

WET MOUNTING
A wet glue method may also be selected for mounting assembled puzzles. Invert completed puzzle as above, apply adhesive with brush or roller to substrate and align onto back of puzzle. Invert to face up position and weight for appropriate 4-24 hour drying time. Use of a cold vacuum frame will expedite the bonding process and increase longevity of the initial bond.

PRESSURE-SENSITIVE MOUNTING
Pressure-sensitive boards or P-S films may also be used when mounting paper jigsaw puzzles. Invert puzzle face down, align P-S board, turn upright and weight. As mentioned with wet glues, cold vacuum frames will speed the bonding process and increase long term fusion of pressure-sensitive mountings also.

Laminating over mounted puzzles with either heat-seal films or brush on liquid laminates will further protect and enhance the permanently bonded puzzle, making it better suited to framing without a mat or additional glazing.

BRASS RUBBINGS

Rubbings are made on relatively thin papers that allow the hard wax sticks to conform to and around the shapes on brass nameplates and tombstones echoing the designs and lettering. Since the colored sticks used for the rubbings are hard, high melt wax, it is wise to be careful when dry mounting. Low mounting temperatures between 150-170°F are best if dry mounting methods are to be used. See also page 57 for pressure-sensitive applications.

Orange peel may also result when mounting a thin brass rubbing to a textured substrate. Lower mounting temperature helps cut down on this undesired effect, but a release board, lumpy substrate combination added to warm temperatures and higher pressures during mounting may add up to unwanted visual texture anyway.

MATERIALS
- Brass rubbing
- Breathable, removable, low temperature adhesive
- Smooth substrate of choice
- Tacking iron
- Release paper

1. Set press temperatures and predry materials as needed for mechanical press.
2. Align and tack to substrate.
3. Mount at 170°F in mechanical press 2-4 minutes, vacuum system 4-5 minutes.
4. Remove, and cool under weight.

COLD MOUNTING METHODS

Since the wax sticks are susceptible to higher temperatures, it might be safest to mount brass rubbings using alternative cold methods. Wet and spray adhesives could cockle the thin papers so care should be taken to control the moisture by not over applying the adhesive, and waiting the appropriate open time required for successful mounting (TTPM). If not comfortable with standard wet and spray procedures, this is not a good project to experiment with.

Pressure-sensitive boards and film applications are highly recommended for their dry approach, cold application, and ease of use. Apply to back of the rubbing if using a film, or align the rubbing directly onto the board, cover with release liner and burnish in place with squeegee, rubber roller, or cold frame. Remember the thin paper of the artwork will allow greater orange peel to show when a lumpy substrate and high degree of pressure is used during application. Pay attention to selection of appropriate materials.

PHOTOGRAPHS
RC PHOTOGRAPHS
A resin-coated (RC) photograph is a sandwich of resin coating on either side of a paper core with a developing emulsion on top. They may be wet, spray, pressure-sensitive or dry mounted. Photographs tolerate heat very well though may pickup the contour of the substrate creating what is known as *orange peel*. This is the uneven lumpy surface distortion of the photo relaxed to contour to the base. Smoother mounting surfaces and two-sided release paper or clear release Mylar/polyester rather than top release boards consistently give a better dry mounting result.

Photo emulsion
Polyethylene coating
Paper core
Polyethylene coating

Though alternative application using spray or pressure-sensitive may eliminate much of the orange peel, the resin nature of the nonabsorbent backing lessens the permanency of the bond.

MATERIALS
    RC Photograph
    Permanent, breathable, tissue adhesive
    Rigid substrate
        Smooth wood/metal/glass/hardboard/MDF
    Tacking iron
    Release materials

1. Predry as needed for mechanical presses.
2. Z-method tack photo to adhesive to substrate.
3. Assemble mount package:
       Two-sided or Mylar release sheet
       Plastic foil overlay or ColorMount cover sheet
       RC Photograph
       Permanent adhesive
       Substrate
       Release paper
4. Mount at 185°F, 1-3 minutes mechanical; 3-5 minutes vacuum press.
5. Remove and cool under weight.

NOTE: Hot Press Overlay Foil-Acetate Film is a 1.5 mil acrylic sheet with no silicone release. When layered over a high gloss photo it will protect the photo from silicone damage during mounting. ColorMount cover sheets are polyester encapsulated paper with no silicone release to react with the photo surface emulsions, though there is orange peel.

OVERSIZED PHOTOGRAPHS: *MECHANICAL PRESSES*

Mounting oversized photos varies whether in a dry mount press or vacuum press due to the nonporous nature of a RC photo. Mechanical presses create the mounting pressure as soon as the press is clamped into the locked position, therefore any air trapped beneath a large photo (totally fitting within the confines of a single mounting) is removed prior to heating up the mounting package.

Mechanical press procedure and materials for mounting an 18x24" photo include:
- a breathable/porous RC approved adhesive
- the lowest indicated temperature for adhesive (T)
- 2-3 sheets release paper on top rather than release board
- recommended 2 minutes to mount, varies with size and
- thickness of substrate (T)
- adjusted pressure for selected substrate (P)
- predry materials prior to mounting (M)

MATERIALS
      18x24" RC photograph
      Nonbuffered, cotton fiber, photo board
      Permanent, porous, tissue adhesive
      Tacking iron
      Single-sided or polyester release materials

1. Predry as needed.
2. Tack into position using Z-method.
3. Stack mounting between standard release materials.
4. Set press and mount at 185°F, 2-3 minutes.
5. Cool under weight.

➤ **TIP:** *RELEASE BOARDS AND ORANGE PEEL*
Commercial manufactured release boards begin with an orange peel appearance, being made from paper mountboard. This translates to the surface of a RC photo increasing the orange peel potential if a lumpy substrate is also chosen. Select double-sided release paper, polyester overlay foil, or make release boards with one-sided paper and permanent adhesive on a smoother baseboard.

➤ **TIP:** *IDENTIFYING RC PHOTOS*
Though not always identified by it, if a manufacturer's name is printed on the back of the paper it is most likely an RC or digital photo.

OVERSIZED PHOTOS: *HOT VACUUM PRESS*

The operation of a vacuum system integrates a 45 second to 1 minute delay of the desired mounting pressure while the vacuum is being pulled. Any tissue can begin to activate during that time allowing for the outer edges of the adhesive to set prior to the air being compressed from between the photo and the substrate.

Since permanent adhesives are generally used for mounting photos, the bond is created in the press as the adhesive and materials reach temperature. The domed shape of the photo created by the mounted outer edges will retard the press from totally squishing the air or shifting the outer photo edges necessary to create a flat bond.

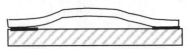

Trapped air can occur when the adhesive sets before the pull of the vacuum

This rarely occurs with small photos or nonbreathables because the air manages to more easily reach the edges before mounting to escape entrapment. Sometimes by the addition of 2-3 sheets of top release paper, heat transfer is slowed allowing additional time for air to escape before bonding.

Suggested mounting procedure when working with photos up to 18x24" in a vacuum press includes:
- a breathable RC approved adhesive
- the lowest indicated temperature for adhesive (T)
- 2-3 sheets release paper rather than board
- recommended 4 minutes, varies with size (T)
- pressure (P) and predrying (M) are automatic

> ***TIP:*** *HIGH GLOSS FINISHES ON RC PHOTOS*
> Hot Press Overlay Foil (or Seal Graphics Gloss Release Film) is an uncoated acrylic sheet used as a cover layer when mounting high gloss items to help prevent blotching and silicone damage.

> ***TIP:*** *AIR BUBBLES AND THEIR REMOVAL*
> Once an air bubble has been created, no amount of additional press time will flatten it out. Punching an air hole in the substrate will still not allow the photo to shift out, flatten against the board, and eliminate the bubble.

TWO-STEP METHOD

In order to prevent undesired bubbles; use a two-step temperature process of dry mounting. By beginning at a temperature too low for the adhesive to activate, the vacuum is allowed to be drawn and the air to be compressed from within the press prior to adhesive activation. An additional 10 minutes of time is required for the press to draw up to the required temperature, and then mount for the routine 4-5 minutes, for a total of about 15 minutes for a 40x60" press.

When mounting oversized photos larger than 18x24", the suggested two-step mounting method will include:

- a <u>breathable</u> suitable tissue which mounts 185-190°F.
- <u>2-3 top sheets</u> of single or double release paper
- <u>begin at 20°F below</u> normal adhesive (160°F) temperature to squish out air (P)
- <u>turn press up 20°F</u> (180°F) to mount (T)
- once press achieves desired temperature, allow additional 2-3 minutes for actual mounting <u>approximately 10-15 minutes total</u> (T)

By using more than one release sheet on top of a photograph, the platen is slower to heat up and activate the top materials. This allows additional time for the vacuum to be pulled and all air to be removed prior to final mounting.

MATERIALS

    18x24" RC photograph or larger
    Rigid, nonbuffered, acid-free photo board
    185°F permanent, porous, tissue adhesive
    Tacking iron
    Single-sided or polyester release materials

- No predrying is necessary when vacuum mounting.
- Tack photo to substrate using Z-method.
- Begin with press set at 160°F or 20-25 degrees below suggested adhesive temperature, too low to activate the selected 185°F adhesive.
- Place project in press, close, and lock.
- Set to required 185°F temperature for 15 minutes.
- Cool under weight.

High temperatures may create white damage marks on the surface of high gloss photos. This may be the effect of silicone residue from the release materials leaving a slight film. Laminating or photo sprays will cover or camouflage this effect.

FIBRE-BASE PHOTOGRAPHS (SILVER GELATIN PRINTS)

Fibre-base photographs are silver gelatin based emulsion developed on a solid paper core and backing. They have no resin coating on the back side allowing for a permanent bond between the photo paper and the substrate. It is somewhat porous and can therefore tolerate most types of mounting and adhesive.

If the project is an antique or remains the only copy, it should be treated as an original and not dry mounted. Store using conservation methods for storage of the original and use a duplicate for display framing is best. If it is to be mounted, low temperature, removable, porous tissue adhesives are best.

MATERIALS
- Fibre-base photograph
- Substrate of choice (possible 150°F *Speed*Mount HA board)
- Breathable, low temperature, removable, tissue adhesive
- Tacking iron
- Release materials

- Predry as needed.
- Tack photo to substrate using Z-method, or select alternate of float or premounting.
- Position in release materials.
- Mount at 165°F, 1-3 minutes or suggested time for chosen adhesive.
- Cool under weight.

If a vintage fibre-base studio portrait is the last in the series it is considered an original and must be framed as one. If duplicates are desired, they may be inexpensively copied into glossy B/W reprints, but will not have the aged color or texture of the original. A professional portrait studio might be able to exactly duplicate the antique grays and yellowed whites of the aging of the natural portrait when printed as a color rather than black-and-white photo. Reprints will run $30-50.00 rather than $5.00 for glossy B/W.

Become familiar with local photo labs and their available services. Offer duplication or copy of the original as additional framing service, for a fee. Maintain control of the sale by handling as much of the process as possible for the customer.

> ***TIP: ARCHIVALLY-NAMED ADHESIVES***
> Removable acid-free adhesives are not considered conservation, in spite of their name. Since some of the adhesive will always remain saturated into the backing paper the photo will never be in its original state again.

ILFOCHROME CLASSIC (CIBACHROME)
Dry mounting should not be used in conjunction with Cibachrome polyester photos. The dignity of their smooth, glass-like surface and high gloss finish must be retained during framing. Heat will not physically melt an Ilfochrome Classic until it reaches temperatures in excess of 300°F. High temperatures of 200°F will badly damage the shiny surface by blotching it during exposure. Even low 165°F temperatures will contour the polyester into the highs and lows of the selected substrate creating orange peel.

Large Cibachrome prints over 16x20" have a tendency to sag by the weight and soft nature of the polyester film. Mylar corners are somewhat insecure, large photos can buckle or sag from within them. Even archival edge strips will not prevent sagging in the center. Japanese hinges and cooked starch pastes might cause any excess moisture to buckle the connection sights. Acid-free, pressure-sensitive tapes as pendant hinges or better a flange hinge clear across the top will nicely support the photo. Since polyester does not absorb anything, it is also a preservation method. See Ilfochrome Classic: Static Mounting, page 66.

NOTE: RC Cibachromes are not 100% polyester, but rather the same as any paper core traditional RC photograph.

HEAT-SENSITIVES
Sometimes it is known when an item, paper, or paint is heat-sensitive, sometimes not. If there is any doubt, <u>do not heat mount</u>. Perhaps the art should not be permanently mounted using any method at all, but rather preservation hinged.

DIGITALS
There are numerous variables when dealing with digitals from types of inks and papers to printing processes, some are heat-sensitive some tolerate low temperatures of 150°F. Heat mounting is a thermographic risk while wet and spray glues may impact the water soluble nature of the inks. Pressure-sensitive mounting and hinging is recommended. See <u>Heat-sensitive Items</u> pages 60-66, Chapter 8. Also Appendix: pages 145-151.

PAPYRUS, VELLUM AND PARCHMENT
Painted papyrus paper art is not generally heat-sensitive. The reed paper and the paint tolerate heat enough to dry mount, but the dignity of the art suggests it should not be mounted. The natural rippling and cockling of the natural fiber reeds should be allowed to occur through alternative more conservation mounting methods.

The same is true with animal skins. Real sheepskin (called parchment) and real vellum (which is all other animal skins besides sheep) may tolerate heat, but are visually best presented when archivally handled. This allows the skins to expand and contract with relative humidity and temperature variations as in nature. This truly maintains the dignity of the art. Perhaps not flat, but in this case it is the nature of the beast.

COLOR COPIES

In a 1996 test conducted for heat tolerances at 160°F, 190°F, and 220°F, it was determined that at low temperatures *some* of the copies tolerate heat enough to be dry mounted, particularly Xerox (see pages 150-151). In order to be mounted safely the origin of the copy must be known. Since electrophotographic images are made using dry pigmented toner that is set through heated fuser rollers, any later application of heat impacts the even gloss finish. If the toner setting temperature is lower than the desired mounting temperature, toner gloss damage will be the result. Gloss damage appears as a blotchy or mottled look. Cold mounting P-S or hinging methods are recommended. Heavy or uneven saturations of wet and spray glues may cockle thin copier papers.

DIGITAL IMAGES

In 2001, a series of research and heat tolerance testing was completed on other digital images including: electrostatic, thermal transfer (including dye sublimation), and inkjet (including phase change, piezo, thermal and continuous flow). Although many of the digital technologies tolerate 150°F allowing them to be dry mounted with Nielsen Bainbridge *Speed*Mount, the origin of the digital should still be known prior to applying heat. See Digitals pages 60-64, and Appendix, pages 145-149 for more explanation, discussion, testing procedures, and charted results. As with the above color copies, without the known origin of the image either a second copy for testing or P-S mounting is recommended.

GICLEES

A giclee is a common term used for fine art digital prints, particularly ones printed with continuous flow printers. There are two levels of giclee currently being printed: fine art giclee, and decor-quality giclees. The Giclee Printers Association (GPA) has attempted to set acceptable standards to keep these inkjet prints separated. Fine art giclees are usually signed and numbered and therefore should be framed to preservation standards, including traditional conservation/preservation mounting.

DIGITAL PHOTOGRAPHS

Any photograph electronically recorded, manipulated and printed with an inkjet or thermal transfer printer can be heat-sensitive. These images highly resemble traditional RC photos. To help identify digital photos check manufacturer identification (i.e.: Kodak Professional Paper, Fuji...) on the verso side of the photo for the words *digital image* or *electronic imaging* which may or may not be there. Digital photos are made up of tiny dots of ink that are between 300-2400dpi, detectable only with a loupe. Digitally produced photos are pixilated, meaning they appear as tiny squares of color when greatly enlarged on a computer screen. Traditional RC photos are solid continuous tone color. Any traditional photo (slide, negative, print) that has been scanned for computer correction or manipulation will become a pixilated image. Any photo then printed out will be considered a digital photo regardless of origin. Since digital photos may be heat-sensitive, alternative methods for mounting should be considered.

PART TWO

Laminating

11

Laminating Basics

WHAT IS LAMINATING
Laminating is a process of applying any durable clear pressure-sensitive or heat-seal film to a flat surface for the purposes of protecting and enhancing. Films may be acrylic, polypropylene, polyester or vinyl depending upon the equipment used, purpose for the lamination, and whether it is to be cold laminated with liquid laminates, roller machines or heat set in a press. For additional information on cold laminating in the commercial market, see Chapter 8.

In the picture framing industry, laminating is the heat application of a protective vinyl laminating film to the surface of paper art or photograph as a glazing substitute. It is a nonreversible alternative to glass, which is washable, durable, permanent, lightweight, nonbreakable, will not fingerprint, has some UV protective properties, and is nonporous. Films come in an assortment of finishes and textures and all but one are made of vinyl materials. There is one high gloss finish made of polyester in which the technical mounting procedure varies slightly due to the variation in material composition.

SURFACE LAMINATING VS ENCAPSULATING
When only one side of a project is to be coated or covered with a protective film, it is known as over- or surface-lamination. Enclosing or sealing an item between two sheets of clear film is known as encapsulation. There are conservation and nonconservation methods of encapsulation. The conservation method is the process designed to protect documents from destructive outside elements and is totally reversible, while laminating encapsulation with heat-seal polyester films is not.

Laminating encapsulation directly relates to the use of permanently fusing laminating film whether vinyl or polyester to both sides of a paper or document using heat-seal or cold P-S methods. Menus, ID cards and catalogue pages are often encapsulated two sides to protect them from moisture and damage.

One-sided laminating using either vinyl or polyester film is known as over-laminating or surface laminating. It may be achieved in heat presses with tissue adhesives or with a roller laminator using pressure-sensitive adhesives for commercial markets.

MARKETING POTENTIAL

Laminating is a natural progression from dry mounting within the framing industry. No additional equipment is required and the only additional materials for heat-seal surface laminating include:

    Vinyl laminating film
    Overlay foam or foam plastic
    OPTIONAL: Perforator or piercing tool

Marketing advantages of surface laminating include:

- Unbreakable glazing means the ability to offer framed art to nursing homes, hospitals, day-care centers, preschools, children's rooms, pediatricians' offices, detention facilities or anywhere glass is not allowed for safety reasons.

- Moisture proof vinyls mean they may be used in high humidity areas such as public swimming pools, bathrooms, kitchens, boats, and makes them washable surfaces. As framed art, signage, or charts allow for use with washable markers.

- Though available in many finishes, the nonglare nature allows hanging directly under light or opposite windows.

- Lightweight vinyl cuts down on the general weight of glazing for shipping and framing of oversized charts, maps, and artwork.

- Ultraviolet protection is not promoted as a marketing element, but the laminating films inhibit UV rays from penetrating, reducing fading from natural sunlight, fluorescent, and tungsten bulbs.

PRICING

Pricing is simple. Use the same price chart currently used for mounting since the process is basically the same. Films run slightly higher in cost than mounting tissues but the mounting substrate is included with the mounting price. Laminating prices need not reflect substrate costs, thus offsetting the difference. It is not necessarily cheaper than glass; it is a glazing alternative.

POLYESTER FILMS
TWO-SIDED ENCAPSULATION

The polyester films developed for encapsulation were originally developed in the early 60s for the protection of paper rather than to make paper look good. Films are available in a number of varying thicknesses. The 1.5mil film differs in composition from the 2, 5, 10 and 15mil in adhesive, application and designated use.

The thinner 1.5 mil polyester is the more inexpensive, economical grade of film designed to be used most specifically with paper and ink. The low density polyethylene adhesive requires a relatively high mounting temperature of 230°F -275°F. Since the adhesive will not fuse to photographic emulsion it is restricted to use on nonphotographic papers only. It can be written on, is water repellent, durable and comes in gloss and matte finishes.

Polyester and polyethylene films cannot be perforated for use over nonporous items because the film will not heal in the press during mounting.

The thicker films (2, 5, 7, 10, 20 mil...) are considered a more commercial grade designed for use with photographs, toner copiers and other special applications. The adhesive is a modified copolymer which sets at a lower mounting temperature of 220-240°F, which is why it is more photo friendly. Thicker films work well for free standing displays, place mats, ID cards, menus and other two-sided encapsulation needs. Polyester films have an adhesive side which is best identified by its dull appearance, are not tacky to the touch, do not have a release paper backing, and therefore are not repositionable during setup as are vinyl films.

Polyester encapsulation within a heat press is not advised because of the potential for trapped air. Films do not heat when perforated and there is no way to ensure total removal of the air from between the two sheets during mounting.

TIME
The time required in a press runs 3-10 minutes depending upon the size of the project and the substrate.

TEMPERATURE
Polyester film is a tough material, which sets in a heat press between 230-275°F. The two-step temperature process of 180°F turned up to 275°F may be required when a vacuum press is used.

PRESSURE
In a mechanical press, shims may be needed for the lack of substrate to adjust the 45-degree handle. In a vacuum press, the two-step temperature method allows the vacuum to draw the air out creating pressure before the adhesive melts.

MOISTURE
Predrying papers will prevent condensation of trapped moisture within encapsulated projects when using vinyls. Polyester encapsulation is not recommended for framing presses.

ROLLER LAMINATORS

The entire encapsulation process involves the use of nonporous polyester films. Since polyesters do not breathe, they were originally designated for use with roller laminators designed to squeeze the air from between the layers as heated rollers apply pressure and fuse the film sheets together at the same time.

Heat-seal office laminators use polyester pouches and were predominantly developed for specialized use within schools, libraries, print shops, graphics, and reprographics houses. Large-scale encapsulation operations require purchase of a commercial roller machine. Current trends in computer graphics, inkjet applications in advertising, and expanding markets indicated new laminating processes for roller laminators, both hot and cold, are rapidly developing. Applications for exterior signs, point of purchase displays, banners, and floor graphics in the digital market are thriving, and with the advent of collectible heat-sensitive digital images, rollers are now a logical addition to almost any custom framing equipment.

High tack, pressure-sensitive, vinyl films designed specifically for use with rollers are available in numerous finishes including matte, satin matte, gloss and hi-gloss, along with assorted textures of canvas, linen, emery, sand, crush, and leather.

VINYL FILMS
FOR SURFACE LAMINATION

In the early 80s, vinyl laminating film was developed for use specifically within the picture framing industry. Single-sided application of a laminate is known as surface- or over-lamination, and these films are used as a glass substitute.

Vinyl films were developed specifically for use within a heat press rather than a roller laminator, and feature a removable release paper backing. Laminating films are repositionable when initially applied to the poster art prior to mounting. When left in position on the surface of a poster unmounted for any length of time, peeling the film sheet from the face of the poster may lift off some of the ink with it.

TIME
Press time averages 5-15 minutes depending on the procedure, size, and thickness of the project.

TEMPERATURE
Vinyl film is an easy to apply material, which sets between 185-225°F depending on the manufacturer.

PRESSURE
Check mechanical press adjustments for all selected materials. Vacuum systems require the two-step process to ensure pressure is applied prior to bonding.

MOISTURE
Predrying artwork prevents condensation and trapped moisture. Perforation is required with nonbreathables.

Vinyl films may be perforated to allow for laminating of nonporous photographs without trapping air. Because of lower mounting temperatures between 180-225°F they may also be used with foam boards that begin to melt down at hotter temperatures of 230°F.

There are a variety of finishes and textures available for surface laminating, in custom framing, from many different companies. All vinyl films for framing require use of overlay foam or foam plastic to ensure proper adhesion and desired texture during mounting.

OVERVIEW OF POLYESTER LAMINATES
- Developed in the 60s for the protection of paper
- Designed for use with roller/laminator machines in the commercial market
- Have no release paper liner
- Nontacky and slippery to handle
- Does not require overlay foam
- Cannot be perforated for use over nonbreathables
- Mounts at high temperatures of 240-275°F
- Curls near heat
- Sold in a wide variety of mil thicknesses
- Available in rolls

OVERVIEW OF VINYL LAMINATES
- Developed in the 80s as a glazing substitute
- Designed for use with heat presses in the framing market
- Have a release paper liner
- Tacky and repositionable prior to mounting
- Requires overlay foam
- Applicable over nonbreathable materials when proper perforation techniques are implemented
- Mount at temperatures between 185-225°F
- Usable with a foam board substrates
- High gloss over-laminates, camouflaging as vinyl's with a release liner, are indeed polyester and may not be perforated
- Not marketed to framers in mil thicknesses
- Available in sheets, rolls, perforated (pierced) and nonperforated
- Available in assorted matte, gloss and high gloss finishes and various canvas, linen and emery textures

ADDITIONAL TOOLS AND MATERIALS
The only additional materials required to begin laminating include the vinyl film and overlay foam, also called an overlay blanket, and perforator which is optional for use with nonporous photos.

OVERLAY FOAMS
Overlay foam must be used with all vinyl films to achieve manufacturers desired results, and must cover the entire project being laminated. There are three basic reasons for needing overlay foam during lamination. First, it is designed to create even pressure between the platen and the uneven surface of the films, particularly the textured linen and canvas finishes. Second, it slows down the heating time of the films so that, third, it can better allow for air to escape from the center of the project to the outer edges during bonding. Do not substitute unfamiliar commercial foams for manufacturers overlay foam. Some foam may adhere to the laminating films during the mounting process.

FOAM PLASTIC
Standard overlay foam is ¼" thick, *foam plastic* is the term for ½" thick foam for use in some vacuum presses. The commercial version is a yellow-white color and is used to slow the transfer of heat from the platen to the adhesive. This allows the air and moisture the necessary additional time to be evacuated before bonding occurs. Using ½" foam plastic could eliminate the need for the two-step temperature process of laminating in vacuum presses with rubber impregnated diaphragms that draw quicker vacuums.

FOAM TROUBLESHOOTING
Using damaged foam or not covering the entire project completely will result in an uneven surface appearance.

If the film does not cover the entire adhesive, the foam will adhere to the adhesive when mounted. This will cause no damage to the poster since the foam residue will be stuck to the exposed adhesive on the exterior edge meant to be trimmed away, but the foam will be damaged and unusable for another project of the same size.

PERFORATORS
A perforator or piercing tool is a five-wheel roller designed to punch tiny surface holes in the laminating vinyl to allow for air or steam to escape prior to and during final laminating.

American
cast aluminum
peforator

European
piercing tool

12

Laminating Applications

BASIC TECHNIQUES
Laminating applications vary depending upon the manufacturer and press selected. The heat-seal laminating film is a thin vinyl, which sets in a heat press at a temperature of between 190-225°F. The time required in the same press runs 5-10 minutes depending upon the brand of laminate, overall size of the project, and thickness of the substrate. Basic set-up and preparation for applying films remains fairly constant between manufacturers, so learning the basics will follow through.

Anytime laminating films come near heat the vinyls immediately begin reacting to the temperature with the development of numerous little bubbles. DO NOT attempt to peel up the film after it has begun to react to the heat because there will most likely be ink removal creating irreversible damage to the poster.

STANDARD ALIGNMENT, PREPARATION AND SET-UP
- Understanding the standard procedures for proper laminating will eliminate most potential problems. Pay attention to maintaining a clean area while positioning all films. The static electricity created during removal of the liner can attract unwanted dirt and dust particles. A clean work area when laminating will create the same clean end product as when mounting.
Clean area . . . clean process (see Chapter 3).

- Once a poster has been mounted and the film cut to size, peel back the first few inches of the release paper to expose the tacky adhesive of the film for positioning on the poster. The larger the print or poster being laminated the wider the strip of film should be folded back.

- Line up the bottom of the film with the edges of the poster making certain all adhesive is covered and the film is properly aligned.

- Slide your hand lightly across the surface of the film from bottom to top and press the tacky film into place. There is no need to rub or burnish hard. A light-feathered fingertip check is best.

- Reach under the film, grasp the release paper, and with two hands pull it towards the bottom end of the board.

- Cover with overlay foam and place in press for designated time and temperature. The foam is required to apply even pressure to the surface of the film, slow down the heat transfer, and allow air time to escape out the edge.

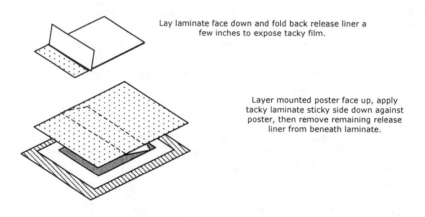

Lay laminate face down and fold back release liner a few inches to expose tacky film.

Layer mounted poster face up, apply tacky laminate sticky side down against poster, then remove remaining release liner from beneath laminate.

LAMINATING PACKAGE DIAGRAMS

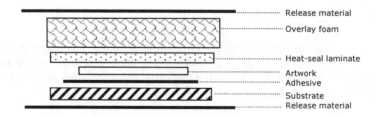

SURFACE LAMINATING OVER BREATHABLES

Floating of art and laminating film in the center of the substrate for laminating will allow for less time wasted with the alignment process during preparation. This way, if foam board is to be used as the base, there will be no crushed or compressed edges once the poster art is trimmed to size.

MATERIALS
- Porous poster prints
- Substrate of choice
- Breathable, permanent, tissue adhesive
- Surface vinyl laminating film
- Overlay foam or foam plastic
- Tacking iron
- Release materials

Standard laminating requires two visits into the heat press. First to mount, then to laminate. Any project may be placed back into the press for additional mounting time, just as when dry mounting, but in order to add time it must first heat the initial time plus the added time. Original time + added time.

MOUNT
1. Predry as needed, for a mechanical press.
2. Tack and mount print in 185°F for 2-4 minutes.

LAMINATE
3. Follow BASIC PROCEDURE from previous page.
   Fold back top edge of release backing, position over mounted print, pull remainder of backing paper from behind.
4. Press temperatures vary 185-225°F depending on manufacturer, turn press up or adjust temperature for chosen materials.
5. Mounting package from top to bottom:
   - Release material
   - Overlay foam or foam plastic
   - Surface laminating film (vinyl)
   - Print (face up)
   - Adhesive
   - Substrate
   - Release paper
6. Place in press 5-7 minutes.
7. Remove and cool under weight.

ONE-STEP MOUNTING AND LAMINATING

Though pricing structures and suggested procedures may indicate a two step mounting process, many art projects may easily be mounted and laminated at the same time. Always tack then apply the laminating film from the same edge so as not to create buckles or potential air bubbles beneath the film. Surface tacking through, both adhesive <u>and</u> print (rather than Z-method) becomes essential to successful, unwrinkled applications. The press must be preset to the required laminating temperature (185-225°F).

MATERIALS

    Porous (breathable) poster print
    Substrate of choice
    Breathable, permanent, tissue adhesive
    Surface vinyl laminating film
    Overlay foam or foam plastic
    Tacking iron
    Release materials

1. Preset press to higher mounting temperature for chosen laminate (if necessary).
2. Cut adhesive should be larger than the print and smaller than the substrate. All laminates must be sized large enough to entirely cover <u>all</u> adhesive tissue.
3. Predry materials as necessary for mechanical presses.
4. Surface tack print and adhesive at center <u>end</u>, not the long side.
5. Fold back inner release liner of film and position exposed film on to the tacked end of the print.
6. Remove remaining release material, pulling from under print. This keeps all exposed sticky adhesive facing the print away from the open air, which helps prevent static from attracting dust and dirt particles during alignment.
7. Assemble mounting package as above.
8. Mount at 185-225°F for 5-7 minutes, remove, and cool under weight.

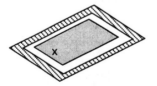

Surface tack print to end of board.

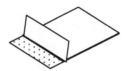

Fold back release liner.

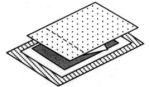

Align tacky side of exposed laminate to tacked end of poster/adhesive/substrate unit. Gently pull away release liner by reaching under laminate and drawing toward lower end of board.

HANDLING OVERSIZED ART

Large art requires a great deal of additional working area when dealing with laminating materials. Removing the release paper liner in a clean well lit area with little added clutter will allow for better control of dirt and particles. Static electricity is at an all time high when peeling the paper from the verso side of a large piece.

Do not pull the release paper from the center only as the film has a tendency to trough creating permanent damage creases in the film surface. Another way the release paper might be removed is to roll the release paper onto a cardboard tube as pulling it from the vinyl, much like removing backing paper from acrylic sheeting.

LAMINATING OVERSIZED POSTERS

When preparing to laminate an oversized poster or photo the artwork should be mounted first rather than attempting to do it all in one single step. This will prevent the possibility if anything slipping during the process. Larger pieces are bulkier and take significantly greater care to prevent problems. Mount the poster with breathable tissue adhesive just a little larger than the poster then cut the laminating film just larger than the adhesive, but still smaller than the substrate. Laminating may be done in smaller bites in a mechanical press just as with mounting.

MATERIALS

        Oversized poster
        Substrate of choice, larger than print
        Permanent, porous, tissue adhesive
        Surface vinyl laminating film
        Overlay foam
        Tacking iron
        Release materials

1. Mount poster at 190°F for 2-3 minutes mechanical press, 3-5 minutes vacuum press.
2. Preset press to higher temperature of 185-225°F and allow to warm up.
3. Fold back inner release backing of film 4-8 inches to accommodate larger print, repositionable adhesive won't hold well enough to the print for removal when only 1 inch is exposed.
4. Position exposed film over print and smooth out.
5. Remove remaining release material, pulling liner from under print and rolling it onto a tube for greater control, rather than by hand.
6. Assemble mounting package, including overlay foam.
7. Place in 185-220°F press, 7-10 minutes.
8. Remove and cool under weight.

PHOTOGRAPHS
PERFORATING FOR NONBREATHABLES

When placed in a heat press for a matter of minutes, a porous poster print is removed as a beautifully mounted and laminated project ready for fitting. Breathable materials beneath any nonbreathable laminating film still allows for air to be forced through porous paper toward the bottom, away from the film. See suffocation diagram, Chapter 2, page 14.

Laminating photographs differs from laminating prints only in that a resin-coated photograph cannot breathe, making it more difficult to remove all the air from between the laminating film and the photo. In some cases, even paper art or posters with heavy nonporous inks will also fall into this category. If there is a fear of porosity, the solution is to begin with a perforated film.

Laminating films come as both vinyl and polyester, but only vinyl films may be perforated or pierced because the holes will melt closed with the application of heat during the mounting process. Films may be purchased preperforated in both precut sheets and rolls, or use a hand held perforator for manual control.

Manual perforation must always be done with the film face up to prevent any paper fuzz from being forced into the lightly tacky film. All perforation holes need to be about ¼" apart for adequate air escape. Any closer is overkill, farther apart might allow for trapped air. The base weight of the American perforator is generally adequate to make large enough holes for the air to escape while small enough to easily reseal during mounting.

MANUAL PERFORATION

- Lay vinyl film face up on mat board or self-healing mat. Do not perforate on foam board OR the substrate to be used. Foam is too soft, perforations will damage the substrate, allowing the texture of the holes to transfer to the mounted image, and glass is too hard and will not allow for penetration.
- Using a perforating or piercing tool, allow the weight of the tool to create the holes in the film.
- Guide the tool around the film surface in all directions so the holes are an average of ¼" apart.
- Set tool aside and proceed with the laminating process.

Roll perforator in all directions so that holes end up ¼" apart.

LAMINATING PHOTOS: *MECHANICAL PRESSES*

When placing a nonbreathable laminating film over another nonbreathable, such as an RC photograph, the film must be perforated to allow air to be compressed from between the two nonbreathable surfaces prior to the mounting of the film.

MATERIALS
      Nonbreathable photograph
      Substrate of choice
      Breathable, permanent, tissue adhesive
      Surface vinyl laminating film <u>and perforator</u> OR
        commercial preperforated laminating film
      Overlay foam or foam plastic
      Tacking iron
      Release materials

1. Predry as necessary for mechanical press.
2. Mount photo at 185°F, 1-3 minutes.
3. Turn press to higher laminating temperature 220°F, if necessary.
4. Apply perforated film as on page 118, steps 5-7.

There is no danger of trapped air being sealed under a perforated film sheet when using either a hardbed or a softbed mechanical press. All air is compressed immediately upon press closure before materials begin to heat up.

Alternately, in a vacuum press, there is a 15 second to 1 minute delay in the application of the pressure while air is being drawn from within the sealed press. Prior to the draw of the vacuum, all materials are beginning to warm up and could begin to set prior to the air being compressed from between the nonporous layers, leaving air bubbles. A two-step process of mounting is recommended.

Some lower temperature laminates used with thicker ½" foam plastic and shallower distances between diaphragm and platen may not require two-stepping. Check with suggested manufacturers procedures for each individual press and the materials selected.

> ➤ ***TIP:** SUFFOCATION OF NONPOROUS LAYERS*
> Suffocation will kill any potential mounting or laminating project. Only one layer in any project may be nonporous. If more than two layers in a project are nonporous, unwanted air could become trapped within the mounting.

TWO-STEP PROCESS FOR LAMINATING PHOTOS: *VACUUM PRESS*

The vacuum press, having delayed pressure due to the slower draw of the vacuum sometimes requires a two-step laminating process for predictable, worry free results. Begin at a 25-30 degree lower temperature than recommended for laminating, place project in the press, immediately turn up the press temperature to the required 185-220°F laminating temperature and add time to allow for temperature to pull up.

It will take approximately 10-15 minutes for the press to heat up to desired laminating temperature, allowing the press to pull a vacuum before achieving film-bonding temperature. Beginning with the press temperature set to normal laminate temperatures, the laminate could begin to bond too soon and potentially trap undesired air between the film and photo just as though it were never perforated at all.

MATERIALS
- Nonbreathable (nonporous) photograph
- Substrate of choice
- Breathable, permanent, tissue adhesive
- Perforated laminating film
- Overlay foam
- Tacking iron
- Release materials

1. Begin with press set press at 185°F mounting temperature, or 25-30 degrees below recommended press temperature for chosen laminate.
2. Mount photo to substrate 185°F, 3-5minutes (see Chapter 10, Dry Mounting).
3. Remove and cool artwork under a weight.
4. Fold back inner laminate release liner to expose adhesive, and align perforated film onto mounted project.
5. Peel remaining liner from laminate by pulling from beneath print.
6. Assemble laminating package:
    - Top release paper
    - Overlay foam
    - Mounted print, adhesive, substrate unit
    - Bottom release paper
7. Place in 185°F press and <u>immediately</u> turn press up to 220°F, mount 10-15 minutes or watch temperature gauge. When press reaches 220°F, mount a few minutes more. The project does not require any length of time at the lower temperature; it only needs to begin processing at the low temperature.
8. Remove, and cool under weight.

With all 220°F laminating films, the two-step process is nearly always required to ensure nonsuffocation. Many lower temperature laminates (185°F) may not require this process.

CANVAS TRANSFERRING
COPYRIGHT
Depending upon legal precedents set in individual states, currently it appears canvas transferring may only be offered as a service, and one may break copyright law by purchasing images specifically to alter for sale. Transferring studio portraiture without the permission of the photographer (not the consumer) is not legal. The image is a print; the original negative remains owned by the photographer. The best way to approach a photographic transfer from a professional studio is by offering it as a service to the studio directly or establish an open agreement concerning altering their images for customers.

Open edition poster prints are not necessarily free from copyright. If the publisher offers images both on paper stock and canvas, the transfer must be purchased from the publisher directly. Only then will both the publisher and artist be in agreement over the variation of the original image, and both will receive appropriate royalties.

Despite copyright laws, it has become increasingly difficult to prevent illegal infringement. Photocopiers and computer scanners have made it easy to duplicate and enlarge pages from books and catalogues, encouraging creative framing practices by transferring them to canvas. Pay attention to the laws and read all trade magazines to stay updated. Copyright infringement is an illegal practice and ignorance is no excuse.

EQUIPMENT AND MATERIAL VARIATIONS
Manufacturers' equipment varies from depth of mounting area inside a vacuum press (1-4") to the material the diaphragm is made of. Laminates run anywhere from 185°F to 220°F as suggested mounting temperatures, and that coupled with the time it takes to draw the vacuum makes for slight application variations within different systems.

General laminating is easy to modify by checking suggested temperature requirements, and times in any press may be increased with little overall impact. Canvas transferring procedures also vary depending on the press, and have slight modifications depending on vacuum suction time and suggested temperature of the selected laminate. Be sure to test any of the included procedures found in this book and adjust the times, temperatures, and any other procedure variances according to your equipment and materials. All transfer may be placed back into any press for additional mounting time if removed too soon. Add the original time and the additional time for the full second mounting cycle.

> ➤ *TIP*: *TEMPERATURE VARIATIONS*
> Laminates vary in suggested temperatures from 185-225°F. Check with the manufacturer for recommended times and temperatures. Larger transfers may take up to 15 minutes per bite.

RC PHOTOGRAPHS TO CANVAS: *MECHANICAL PRESS*
Transferring the surface emulsion of a photograph to canvas is easily achieved with laminates. Remember this is only possible with RC photographs and contemporary fibre-base photos (not antiques). Ilfochromes cannot be split from inner paper because of they are comprised of 100% polyester. Any RC photo, including RC Cibachromes, may be used in this creative application.

MATERIALS
        8x10" RC photograph/color or black and white
        11x14" Perforated laminating film
        11x14" Heat-activated adhesive coated canvas, white or natural
        Overlay foam or foam plastic
        Release material

1. Perforate film and size all materials, predry as necessary.
2. Trap the 8x10" photo centered between 11x14" laminate/release liner creating a self-envelope.

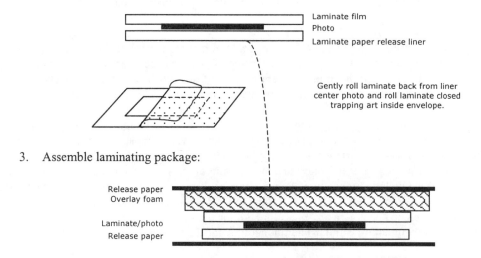

3. Assemble laminating package:

4. Mount laminate to photo 185-220°F, 5 minutes.
5. Remove envelope from press, no need to cool.

6. Place face up on hard surface, begin peeling emulsion/laminate decal from photo paper core, revealing a photo decal. Peel at a 180-degree hairpin back onto itself for maximum paper removal.

 Laminate/photo emulsion
Backing paper

7. Align decal onto HA adhesive coated canvas.
   If stretching onto bars, center an 8x10" on 11x14" canvas.
   If mounting onto 3/16" foam, match canvas to photo size.

8x10" centered
for stretching

11x14" full size
for mounting to 3/16" foam

8. Assemble transfer package:
   Release paper
   Overlay foam
   Laminated photo emulsion on canvas
   Release paper

9. Mount 220°F, 5 minutes, sometimes longer as needed.
10. Completed transfer may be mounted to foam board or stretched onto bars for framing as noted in #7, above.

> ***TIP:*** *RIGID BOARD STIFFENER*
> When mounting in a vacuum press a stiffener needs to be inserted beneath the project to protect the canvas from undue wrinkles in the diaphragm when the vacuum is drawn.
>
> Overlay foam
> Laminate decal on canvas
> Release liner
> Rigid board stffener

RC PHOTOGRAPHS TO CANVAS: *VACUUM PRESS*

The delayed applied pressure of the diaphragm may require a two-step laminate method depending upon make of press and temperature of laminate. The following process is for use with higher temperature 220°F laminates.

MATERIALS
    8x10" RC photograph/color or black-and-white
    11x14" Perforated laminating film
    11x14" HA adhesive coated canvas, either white or natural
    Overlay foam
    12x16" Rigid mount board stiffener (not release board)
    Release material

FACE-UP METHOD

1. Trap photo inside perforated laminate/release liner (page 124).
2. Assemble laminating package:
       Release material
       Overlay foam
       Laminate/photo/release liner envelope
       Release paper
       Rigid board stiffener
3. Mount perforated laminate to photo using two-step process, begin at 185°F, turn press to 220°F for 15 minutes.
4. Remove from press, do not cool, and discard laminate release liner.
5. Begin face up on hard surface, peel photo emulsion/laminate at a 180-degree hairpin back onto itself for maximum paper removal. Do not roll onto dowel or tube, it can pick up too much paper.
6. Align decal onto adhesive coated canvas, centered for stretching.
7. Assemble canvas mount package:
       Release paper
       Overlay foam
       Laminate and photo emulsion onto canvas
       Release paper
       Rigid board stiffener (under release paper, page 125)
8. Mount 220°F, 5 minutes, sometimes longer.
9. Completed transfer may be mounted to foam board or stretched onto bars.

The rigid board stiffener protects the thin nonsubstrate mounting from permanent wrinkles or creases caused by the rubber diaphragm as the vacuum is pulled. **Do not use a silicone release board**, it can suction up to the platen trapping air in the package and reducing the degree of canvas texture.

FACEDOWN METHOD

Presses that pull a quicker vacuum, the result of a narrower drop between diaphragm and platen, also suggest a facedown approach to canvas transferring. The concept allows the sponge foam to press up into the textured canvas during final transfer while the platen substitutes as the rigid stiffener surface.

MATERIALS
    8x10" Color or black-and-white RC photograph
    11x14" Perforated laminating film
    11x14" HA adhesive coated canvas, white or natural
    Overlay foam or foam plastic
    Release material

FACEDOWN

1. Trap photo inside perforated laminate/release liner for initial laminating step.
2. Assemble laminating package (top to bottom):
       Release paper or polyester release film
       Laminate/photo unit, ready for mounting
       Overlay foam or foam plastic
       Release paper or polyester release film.

   > Pay attention to the inversion of the laminate unit and overlay foam.

3. Mount perforated laminate to photo facedown on foam at 185°F, 5 minutes.
4. Remove from press, do not cool, and discard laminate release liner.
5. Place faceup on hard surface, peel emulsion/laminate from paper core, revealing a laminated photo decal. Peel at a 180-degree hairpin back onto itself for maximum paper removal.
6. Align decal onto HA adhesive coated canvas, with or without release liner.
7. Assemble second mounting package with decal aligned onto canvas:
       Polyester release film
       Facedown laminate/photo decal on canvas
       Overlay foam
       Polyester release film
8. Mount 185°F, 5 minutes, sometimes longer.
9. Completed transfer may be mounted to foam board or stretched onto bars.

Canvases prepared for heat transfers come raw or heat-activated; with or without release liners; and/or with a light pressure-sensitive tack to help positioning prior to mounting. They are available in white or natural in various textures and weights. Shop around. Raw canvas material may be used for transfers by preparing them in advance. Liquid vacuum glues that reactivate under heat such as Berto, Sure Mount or VacuGlue 300 may be applied to the canvas, let dry, then completed as above. Generally, white background prints or photos should be applied to a white canvas to preserve the white color. The natural canvas color will tinge the white and appear dirty.

FIBRE-BASE PHOTOGRAPHS TO CANVAS
If the fibre-base photograph is a contemporary print and not a vintage portrait, it may also be transferred to canvas. The antique could technically be transferred but should probably be conservation treated instead. Fibre-base photos are capable of being transferred using either wet or dry methods.

DRY METHOD
The thick paper core allows fibre-base photos to be laminated and stripped as either a RC photo or soaked poster print. The dry stripped method may leave thicker paper remaining on the peeled decal cutting down on the overall canvas texture. Applying the dry decal to canvas does not allow for maximum texture on the transfer. Misting or moistening the paper backing will soften the paper fibers so they more easily push down into the contours of the canvas, creating maximum texture.

WET METHOD
Fibre-base photos are considered nonporous because of the surface emulsion. Laminate the photo according to manufacturers' suggestions using preperforated film. If using a vacuum system, remember to use the rigid support board beneath all nonsubstrate mountings. Mount at 185-220°F for 5-15 minutes depending upon the chosen procedure.

After laminating, remove and discard the release liner and soak the laminated print in water 15 minutes, following standard techniques for transferring prints. Peel soaked photo **face up** on a hard surface, using 180-degree hairpin separation technique. Rolling the moistened laminated photo onto a dowel will pick up additional undesired paper layers, which cuts down on the end product texture as much as mounting to the canvas with a dry paper decal.

The canvas transferring process regardless of whether it is the dry or wet method must be mounted to the HA canvas while moist. This is in direct conflict with regular TTPM teachings, which stresses no moisture should be allowed in any press to ensure proper bonding. Without this moisture the desired maximum canvas texture may be sacrificed.

> ***TIP:** PAPER MOISTENING*
> The moistening technique for softening of paper fibers on the back of a stripped laminated decal may be applied to RC photographs, or fibre-base photographs, and is already part of the regular approach to transferring paper prints.

If uneven, or too much, paper remains on the decal back it will cut down on the texture and may create an inconsistent canvas texture over the entire completed surface.

UNTREATED OR RAW CANVAS

Raw canvas may be coated with heat-activated wet glue to prepare it ahead for transferring. Dry mount films do not work well on raw canvas. There are too many loose layers attempting to mount together to count on them to all remaining flat throughout mounting.

There are too many possibilities for wrinkles and creases.

Though film adhesives will liquefy in heat systems, they have a self-leveling effect. Rather than contouring evenly to the highs and lows of the canvas, it fills in the lows cutting down on the overall canvas texture.

Adhesive leveling out the surface, cutting down on the overall texture.

Adhesive contouring to the highs and lows

By premounting the adhesive film directly to the peeled decal first, and then mounting that fused unit to the textured raw canvas, the contouring will allow for maximum texture.

WATERCOLOR PAPERS

Watercolor papers make terrific substrate surfaces for transfers of open edition watercolor images. Obviously, mounting a watercolor image onto a canvas substrate would appear out of medium. It would need to be mounted to a high quality rough finish watercolor paper to emulate the original. Since HA watercolor paper has not yet been developed, the same principle of premounting the adhesive film first to the decal and then the selected substrate is the procedure for mounting to watercolor paper.
See Premounting, page 84.

Choose a heavy weight 90# to 300# watercolor stock in rough texture. Some cold press papers might have adequate texture, but "hot press" are too smooth. Laminate, soak, peel, and mount to the adhesive. Then mount the decal/adhesive unit to the paper, for process see page 130.

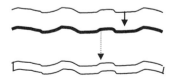

First premount stripped laminated decal to the film adhesive.

Then mount the decal/adhesive unit to the watercolor paper substrate.

POSTER PRINTS TO CANVAS

Transferring a print to canvas varies from photographs in two steps. The laminate does not require to be perforated, unless the inks, paper stock, or finish are nonporous. Inks are sometimes very thick or glossy making them difficult to allow air to pass through. Some low-end clay coated paper stock, usually the most inexpensive posters, also do not allow for air to pass through. These are the ones to watch out for. If there is any doubt at the porosity of the image, perforation is the best solution.

If perforated film is not required, the two step temperature process will not be required when vacuum mounting. As with the fibre-base photo, a print needs to be soaked in water 15 minutes prior to peeling.

MATERIALS
        Porous poster print
        Laminating film (matte or gloss, <u>no textures</u>)
        HA adhesive coated canvas or prepared raw canvas, white or natural
        Overlay foam
        Release materials
        Rigid board stiffener
        Tub or pan of water for soaking

1. Size all materials and set press at proper laminating temperature (185-220°F).
2. Assemble laminating package (top to bottom):

    Release material
    Overlay foam
    Laminate over photo inside liner
    Release paper
    Rigid board stiffener

3. Laminate print with rigid board under bottom release paper, 5 minutes.
4. Remove from press, discard laminate liner sheet.
5. Soak laminated print in water, 15 minutes.
6. Shake off, lay **face up** on hard surface (top of press).
7. Wipe excess water from laminate to smooth out the print,
   it will be held in place by the saturated backing paper of the print.

8. Beginning at a corner in a diagonal movement, roll the laminated print from its paper core backing in a 180-degree hairpin turn. The weight of your hand will easily remove the print from the saturated backing suctioned to the table.

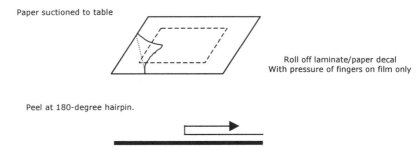

9. Align the decal onto prepared canvas and place in the press for canvas mounting. White background prints should be placed on white treated canvas. Cover all canvas adhesive with surface laminate, so it does not stick to the canvas.
10. Mount 220°F, 5-10 minutes, remove from press and mount to foam or stretch.

> **TIP:** *ADDING BRUSH STROKES*
> Brush strokes may be added for effect using acrylic varnishes such as McDonalds Acrylic Texturing Gel; Grumbacher, Winsor/Newton Acrylic Impasto Paste are all thicker mediums that work well for heavy contours. Some are thick enough for palette knife application while others will need several layers and an artist brush. Acrylic mediums designed for sealing a matte or gloss shine onto paintings are thin and best selected for basic low level brushing. Be sure to match the gloss varnish with gloss laminate and matte with matte finishes.

Crackling and aging techniques of transfers may only be achieved with the use of liquid brush-on laminates and coordinated products onto raw canvas. Soft vinyl laminates stretch rather than crackling.

> **TIP:** *ENCAPSULATING PRINTS*
> Use a roller laminator if polyester encapsulation is desired. If this market has real potential, invest in the proper equipment designed for time and profit efficiency. This would most likely be a small commercial 25-27" roller machine designed specifically for the process.

ENCAPSULATION: TWO-SIDED
POLYESTER FILMS
Polyester films for encapsulation are designed for application by heat-seal roller laminators and may not be successfully used for encapsulation within dry mount presses. Trapping air within the encapsulation will usually occur if attempted in a heat press, either mechanical or vacuum type. Perforation is not an option for avoiding air bubbles, since they will not heal closed when heated. Surface-lamination is easily done with polyesters in a heat press, but not encapsulation.

VINYL FILMS
Since encapsulation in a heat mounting press is not recommended for polyester films, if two-sided encapsulation is required, use the vinyl's designed for dry mounting. The encapsulate will be soft and flexible, but still waterproof and easy to roll for storage or shipping. Creative uses might include placemats, maps for outdoor recreation, or nautical charts. The textured surfaces do not generally make good encapsulates, keep to the simpler smooth finishes.

MATERIALS
    Paper print, fabric, map or chart
    Vinyl laminating film with paper release liner, must be perforated
    Release materials
    Rigid board stiffener

1. Set press to 185-225°F.
2. Predry papers as necessary if using a mechanical press.
3. Cut film larger than paper to be mounted to allow for later trimming.
4. Assemble encapsulating package:
       Release paper
       Overlay foam
       Perforated vinyl laminating film (adhesive down)
       Print, fabric, or chart (face-up)
       Perforate vinyl laminating film (adhesive up)
       Overlay foam
       Release paper, **not release board**
       Rigid board just larger than project
5. Mount 5 minutes, remove, and cool under weight. If mounting in a vacuum system the two-step temperature process is necessary for air removal before setting the vinyl laminate. See procedure found in this chapter.

PART THREE

Appendix

USA/UK TERMINOLOGY EQUIVALENTS

USA	Description/definition	UK
Canvas transferring	Transferring a laminated, stripped poster or photo to canvas fabric for stretching	Canvas bonding
Certified Picture Framer (CPF)	Acknowledgment of excellence awarded by US and UK industry trade associations	Guild Commended Framer (GCF)
Custom frame	Professional picture framer	Bespoke frame
Dust cover	Protective sealer for back of the frame	Sealing tape
Fillet	Decorative liner lip for the inner edge of the window mat or frame	Slip
Filler board	Extra boards behind window mat/mount board inserted before dust cover	Backing board
Gloss release film	Uncoated acetate film for the surface of mountings to maintain high gloss surfaces	Overlay foil
Hardbed press	A manual dry mount press adjusted by use of a wheel	Hardbed press
Laminating film	A vinyl film designed as a glass substitute	Heat-seal film
Matboard	The board cut as a window to surround a print or photo	Mountboard
Mechanical press	A press with a soft foam pad that requires manual adjustment for correct pressure and predrying	Softbed press
Mylar-D	A clear, strong, scratch resistant polyester sheet used for reversible preservation mounting	Melinex 516
Overlay foam ¼"	The foam required during laminating to allow for air transport and even pressure	Foam plastic ½"
Perforating	The punching of tiny holes in vinyl films allowing them the porosity to mount over nonporous posters and photographs	Piercing
Rabbet	Inner edge of frame moulding	Rebate
Solid core board	A mat board with the same color throughout the entire ply	Solid plate board
Support board	Stiff board used to support the mounting when transporting into and out of the press. Often MDF or hardboard	Carrier board
Surface mat decoration	Any combination of lines, and border wash of color, designed to focus on the art	Line and wash decoration
Window mat	The board surrounding the art	Windowmount

FAHRENHEIT TO CELSIUS CONVERSION TABLE											
°F	150	160	170	175	180	185	190	195	200	205	215	225
°C	65	71	76	79	82	85	88	90	93	96	101	107

SUGGESTED MOUNTING METHODS

The following chart is a guideline of recommended techniques only.
It is rarely a question of, can it be mounted?
But rather...should it be mounted at all.

LEGEND
- X   Process approved
- X- Marginal approval at 150°F / Test before use
- +   High tack with Roller Laminator

	Conservation / Preservation	Wet Mounting (by hand)	Spray Mounting (by hand)	Pressure-Sensitive / Film	Pressure-Sensitive / Boards	Cold Vacuum Frame	Dry Mounting	Laminating Tolerance
B/W Photographs	X			X	X	X	X	X
Blueprints (on unbuffered 100% cotton rag)	X			X	X	X		
Brass Rubbings	X			X	X		X-	
Canvas Transfers	X	X				X	X	
Certificates (Replaceable)	X	X	X	X	X	X	X	
Charcoal Drawings	X							
Chromogenic (RC) Photographs	X			X	X	X	X	X
Cibachrome	X							X
Color Copies (4-color)	X			X	X		X-	
Color Laser Prints	X			X	X		X-	
Color Tinting		X	X	X		X	X	
Digitals Images	X			X	X		X-	
Electrophotography	X			X	X			
Electrostatic Printing	X			X	X			
Inkjet on Paper	X			X	X		X	X
Dye Sublimation	X			X	X		X-	
Digital Photographs	X			X	X		X-	
Encapsulated Charts and Maps				+				
Fabric Wrapping		X	X			X	X	
Faxes (Thermographic)	X	X	X	X	X	X		
Fibre-base Photos (Silver Gelatin)	X	X	X	X	X	X	X	X
Fine Art Photographs	X							
Flush Mounting							X	X	X
Giclee'	X							
Ghosting		X	X	X			X	
Heavy Fabrics						X	X	
Holograms	X							
Ilfochrome Classic (Cibachrome)	X			X	X			
Iris Prints	X	X	X	X	X	X		

Cold vacuum frame mounting indicates wet, spray, or pressure-sensitive techniques (as indicated) used specifically in conjunction with a cold vacuum frame, not by hand.

Some procedures require greater pressure during initial bond to better endure more long term bonding.

	Conservation / Preservation	Wet Mounting (by hand)	Spray Mounting (by hand)	Pressure-Sensitive / Film	Pressure-Sensitive / Boards	Cold Vacuum Frame	Dry Mounting	Laminating Tolerance
Japanese (and other) Papercut Art	x	x-	x-					
Laser Prints	x			x	x			
Montage / Collage		x	x	x			x	x
Monoprint	x							
Needleart	x							
Newspaper Text (Two-sided)		x	x	x			x	x
Offset Lithographs (Posters)	x	x	x	x	x	x	x	x
One-sided Text Flier	x	x	x	x	x	x	x	x
Original Art (Pencil / Charcoal / Ink)	x							
Oversized Photographs		x			x	x	x	x
Oversized Poster Prints	x	x		x	x	x	x	x
Papyrus Paintings	x					x	x	
Parchment / Sheepskin	x							
Plain Mounting		x	x	x		x	x	
Polaroid Photographs	x							
Poster Prints	x	x	x	x	x	x	x	x
Premounting				x			x	
Puzzles				x		x	x	
RC (Chromogenic) Photographs	x			x	x	x	x	x
Serigraphs (silkscreen)	x							
Signed Limited Edition Prints	x							
Silks / Sheer Fabrics				x			x	
Silk Embroidery	x							
Thermal Transfers	x			x	x			
Thermographic Tickets	x	x	x	x	x	x		
Textured Decorative Papers							x	x
Translucent Rice Papers							x	x
Vellum Animal Skins	x							
Watercolor Originals	x							
Watercolor Paper							x	x

PHOTOGRAPHIC RECOMMENDATIONS

Many assorted adhesives, substrates and surface mat boards work well with different art and photo prints, but the following recommendations are as stated by the Library of Congress and photo conservators. High humidity makes binders soft and sticky, low humidity will shrink and crack the gelatin binders. Keep humidity between 30-40 % at 68°F, out of direct light, check routinely for insect damage, and fine art images might be best left unmounted.

SUBSTRATES
These are the boards selected for mounting to as a support. They must meet the following standards by being:
- Rigid enough to self-support against a wall without bowing
- Support itself when suspended by hand from one end
- Surface smoothness that minimizes orange peel effect
- Clean cutting density
- Free from long term warpage

MAT BOARD COMPOSITION
Photo board tolerances vary with the photo and the application, whether framed for display or stored. B/W photographs and all RC for archival storage require nonbuffered 100% cotton fiber museum boards. Color RC photos designated for framing display may use mats and substrates of conservation white-core buffered quality, because photos will ultimately fade with time and are subject to natural environmental damage. Micro-chamber technology, though has not stood the test of time, is a plus in this equation.

COLOR RC (a.k.a. CHROMOGENIC) PHOTOS AND DIAZO PRINTS
- ANSI suggests 7.0-8.0 pH with less than 2% calcium carbonate reserve
- 100% cotton fiber when archivally stored
- Tolerates conservation white core when framed as a display photo
- Only frame with conservation when duplicate available
- Traditional paper mat board not acceptable

BLACK-AND-WHITE RC PHOTOS
- ANSI suggested 7.0-9.5 pH with 2% calcium carbonate reserve
- 100% cotton fiber when archivally stored
- Toned prints tolerate conservation boards
- Untoned prints require 100% cotton fiber
- Traditional paper mat board not acceptable

BLACK-AND-WHITE FIBRE-BASE (a.k.a. SILVER GELATIN) PHOTOS
- 100% cotton fiber when archivally stored
- Unaffected by buffering in color conservation boards
- Traditional paper mat board not acceptable

ADHESIVES

There are no current mounting products designed for 100 years with any real knowledge of what occurs to the bond. RC and polyester-base photos are dimensionally stable. The bond is therefore totally subject to stress when humidity naturally expands the mount board and NOT a dimensionally stable photo, potentially allowing for the bond to release.

The suggestions listed here are based on photo conservators and photographic manufacturer recommendations. Half the photo conservators believe in the Ansel Adams school of thought considering the photo print a copy, which is best preserved through dry mounting to protect the edges from damage and the print from bending. The other half of the conservators believe that preservation mounting is the best way to handle long term storage and prevent damage. In either event, both the print and the negative may be considered originals, since the photographer may have touched up or altered the print in question. In many cases, particularly with fine art images, it may be best to select conservation methods for mounting, including sink mats, Japanese hinges, and edge strips.

ADHESIVE SUGGESTIONS
- Permanent, porous, tissue (pH 6.9) adhesive for fibre-base and nonporous polyethylene-coated RC photos
- Pure film adhesives for dry mounting (pH 7.0) are inert and safe for
- RC, polyester, fibre-base, black-and-white
- Acid-free tissue adhesives are alkaline buffered making them less suitable for the slightly acidic nature of RC photos
- Pressure-sensitive adhesives for B/W, polyester-base, Ilfochromes, RC
- No PMA for fibre-base photographs
- No photo spray adhesives for lack of absorption, limiting longevity of bond

OTHER CONSIDERATIONS

Additional tips and considerations with photographs include long term care as well as handling during framing. Care must always be taken with any photos.

- Kodak Film Cleaner as a film solution cleaner
- PEC 12 is a surface emulsion cleaner, NOT for digital photos
- Hot Press Overlay Foil-Acetate Film used as a surface coating layer to protect high gloss photos from silicone damage
- All light is damaging to photos, always use UV glazing
- Keep relative humidity between 30-40% and temperatures at 68°F
- Wavy photos (RC and fibre-base) may be creased if flattened in a press, they are the result of hanging to dry during developing
- Insects and rodents will attack the substrates and binders, check often

FLATTENING PHOTOS

Extreme care must be taken when attempting to flatten any photo. Brittle, aged emulsions can crack if tightly rolled images are forced flat without humidifying them first. Wavy, rippled and curled prints are the result of either the original drying process or years of being rolled while in storage. Badly waved fibre-base photos are the result of hang drying rather than using a print drier during developing. If placed in a dry mount press without flattening permanent creases are inevitable.

HUMIDIFIERS
HORIZONTAL (FLAT)

Using a humidifier will begin to relax the base paper, allowing for gentle waves to be flattened. A developing tray much larger than the photo to be flattened should be filled with warm water and covered with a sheet of clean fiberglass screening across the top. The screen may either be stretched to a strainer or weighted to hold the edges down if temporary. Lay the print on the screen, cover with a sheet of plastic, and weight edges to seal in the moisture. After the moisture has been allowed to penetrate the photo paper place between blotters and weight to dry for 24 hours.

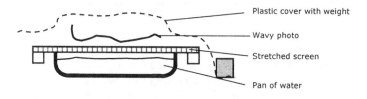

VERTICAL

Tightly rolled prints require more time to absorb and relax the paper base. Place a small open top plastic container or wastebasket into the bottom of a larger plastic container with 2-3 inches of warm water in the bottom. Place the rolled photo into the smaller dry container, close the outer lid and let the photo paper absorb moisture a couple of days. When soft and limp enough, unroll, flatten, and dry between blotters under an even flat weight.

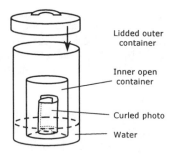

COLD/WET PRESSING

Cold pressing is an aggressive version of flattening an image. This process should only be attempted when the photo is strong, in good condition and there shows no deterioration. This process will not work with RC photos.

- Lightly dampen the back of the photo by misting or with a sponge.
- When limp, lay face down on Pelon over blotter paper.
- Cover with second sheet of Pelon and blotter paper, add weight to dry.
- Prints may be stacked for flattening with the same blotter sandwiches between.
- Change to dry blotters as needed.
- Test the photo back with a drop of water to check for staining before wetting.

Though cold pressing may work with numerous paper based images some artwork is sensitive to moisture such as digital images, original watercolors, charcoal, pencil and ink drawings.

HOT/DRY PRESSING

When flattening in a dry mount press use either blotter papers or 4-ply rag board and release paper to protect the artwork or fibre-base photo.

- Preheat press to maximum of 200°F for photographs and predry all materials.
- Sandwich from top to bottom: release paper, *optional cover sheet (Overlay foil),*
- photo (face up), blotter paper or 4-ply board, release paper
- Close and lock press for 1 minute, cool immediately under a weight.

This process will not insure a flattened image. Badly rippled photos may crease during this process of flattening, which can also occur during dry mounting making hinging recommended. Do not attempt to heat press images that are easily damaged by heat including Ilfochrome Classics, some RC and digital photos, crayon drawings, pastels, embossed surfaces and most digital images.

When attempting to flatten heat tolerant high gloss photos be careful of textured papers, blotters, ColorMount Cover Sheets and release materials that might damage the photo surface. Overlay Foil-Acetate Film should be used to protect the surface of the photo even when only attempting to flatten prior to mounting. The silicone from the release paper, paper fibers from Kraft paper, and texture from Cover Sheets can all still damage the surface.

Think before attempting any corrective technique. Consult a conservator any time the possibilities might be beyond the technical skills of the framer. Damage is often irreparable. Knowing when to consult another professional **is** being a professional.

SELECTING A PRESS

Changes in mounting equipment only occur when equipment needs replacing due to age, growth demands, or a general dissatisfaction with the current system. Time, mess, cleanup, health hazards, artwork risks, are all reasons to upgrade or purchase new equipment.

By asking a series of questions concerning current daily mounting materials, procedures, inventory, and prospectus for future growth the most efficient type of mounting procedure may be targeted to fit specific needs and goals.

> How much wet or spray mounting is done?
> Is mounting subcontracted to another framer?
> Are mounting or laminating jobs turned away?
> Is there heat mounting competition in the immediate area?
> Is there additional marketing potential as a subcontractor for other framers?
> Is there a market for additional mounting growth? Laminating growth?
> Are there potential engineers, architects, day care, real estate, or government offices in the vicinity?

Resale value of heat mounting equipment remains high and mechanical presses seem to live forever, its generally best to purchase something a little larger than exactly what is required at this time. Once heat mounting is selected, the right piece and size of press is determined by asking:

> How much 24x36", 32x40", 40x60" or 4'x8' foam board is bought and used a month?
> Are poster or limited editions sold?
> How many oversized pieces are currently mounted?
> Is inventory increasing into larger images?
> Is photography a viable market?
> What is an average photo size?
> How many oversized photos are mounted?

Think about the targeted market plan five years ahead to help determine appropriate equipment needs. What is lusted after may not always be the best choice for the particular market or inventory.

NOTE: Mounting requirements for wide format printing, as compared to offset printing, is another important consideration. As technology progresses it appears the newest piece of professional custom framing equipment (next to the CMC) will be the cold roller laminator. The high tack pressure-sensitive application using a roller will fill the void when heat mounting is not recommended.

PRESS COMPARISONS/ PROS AND CONS
MECHANICAL PRESSES (SOFTBED AND HARDBED)

PROS
- Mounts and laminates
- Capable of oversized multiple bite mounting
- Takes up less physical table top space
- Uses 110 volt for softbed, 220 volt for hardbed
- Lower capital investment cost

CONS
- Single mounts only up to 24x32" (softbed), 24x24" (hardbed)
- Must be manually adjusted for pressure
- Requires predrying of all mounting materials
- Top may not be also used as a work space
- Requires same surrounding area as substrate size

HOT/COLD VACUUM PRESSES

PROS
- Heat mounts, cold mounts, laminates
- Automatically adjusts to substrate thickness
- Automatically pulls moisture from materials
- Multiple mounting capabilities saves time and $
- Most mounting adhesives may be used (including wet)
- Visual monitoring of mounting projects with glass top units
- Glass top doubles as cutting surface

CONS
- Maximum mounting limited to press frame size
- No oversized multiple bite capabilities
- Monitoring of vacuum pump, filter, oil levels on some models
- Requires 220 volt, three or four-wire hookup and electrician
- Higher electrical usage

There remain manufacturers variations in all presses, techniques, procedures, release materials, overlay foam and order of stacking within the mounting sandwich. Check owner's manuals for specifications.

ADHESIVES

This is a general cross reference chart of commercial adhesives only, not a listing of priority or suggested adhesives. There are many additional private label adhesives, sprays etc. in the above categories, this remains only a sampling for reference. Contact local distributors for additional product information.

DRY MOUNT ADHESIVES
FOTOFLAT (discontinued) / FOTOMOUNT...
    Low temperature 160°F / Removable / Breathable / Tissue-core
    Bonding delicates (silks / watercolors / photos)

MT5/TM-1/POSTERMOUNT...
    Reformulated 200°F / Permanent / Nonbreathable / Tissue-core
    Thin glassine tissue with adhesive
    Bonds paper to paper / NOT for RC photos

COLORMOUNT / TM-2 / TRIMOUNT / SUPER UNIMOUNT / PROMOUNT
    Low temp 180°F / Permanent / Breathable / Tissue
    Same adhesive as MT5 *with porous tissue*
    Tissue allows air to pass through
    Bonds photos / plus all that MT5 does

FUSION 4000 / TM-3 / FLEXMOUNT / ACID-FREE MOUNTING ADHESIVE / VERSAMOUNT
    190°F / Removable / Breathable / Film
    No paper core / 100% neutral
    Bonds fabrics / paper / photos (monitor bleed)

BUFFERMOUNT / TM-4 / DRYCHIVAL / CONSERVABOND / ARCHIVAL TISSUE / SAFEMOUNT
    Low temp 160°F / Removable / Breathable / Tissue / Acid-Free
    Clean easy application / acid-free
    Bonds delicates (silks / vellum / Japanese papers)

HEAT-ACTIVATED MATERIALS
*SPEED*MOUNT / THERMOMOUNT / SINGLE STEP / HARTMOUNT / NUCOR HEAT ACTIVATED...
    150°F-200°F / Permanent and removable / mount boards, foam and canvas
    Convenient for production speed / price reflects adhesive + substrate

PRESSURE-SENSITIVES
PMA / PERFECT MOUNT / GUDY 870...
GATOR / QUICK STIK / BAINBRIDGE or NUCOR SELF-ADHESIVE BOARDS / HARTAC...
    Assorted boards and films from numerous manufacturers
    Permanent bond after 24 hours under weight
    Available as low, medium and high tack

WET GLUES
VACUGLUE 300 / SURE MOUNT / pHABRIC ADHESIVE / LAMIN-ALL / BERTO / UNIMOUNT...
    Removable / dependable depending upon manufacturer / hand, cold and heat applications
    May be brushed / brayered / sprayed
    Clear bonding for ghosting / color tinning

SPRAYS
CORONA VM / SURE-MOUNT / SUPER 77 / VACUMOUNT
    Removable / permanent depending upon application
    Cold vacuum mounting in 30 seconds to 5 minutes
    Clear bonding for ghosting / color tinting

FOUR BASIC DIGITAL PRINT TECHNOLOGIES
ELECTROPHOTOGRAPHY
Dry toner B&W and four-color copies from an existing document
Wet liquid ink photocopying and dry xerographic toner copying. Xerography is Greek for *to write dry*, and is an electrically charged drum that receives an illuminated image that is converted into a dot pattern. It picks up toner, rolls it onto paper and fuses with heat rollers. May or may not be affected by heat and laminate applications.

ELECTROSTATIC PRINTING
Pigmented toner on dielectric paper not used for fine art, laser printers
Uses static electricity to transfer an image to a charged drum. A laser negatively charges a cylinder to the image pattern, positively charged toner is attracted to the negative areas of the drum, special dielectric paper is pressed against the drum to receive the toner and is set through heat rollers. This process uses a heat set ink, not thermal papers.

THERMAL TRANSFER PRINTING
Four-color printers using dyes and pigments on a ribbon of wax-like paper that transfer with heat during printing, also called dye sublimation, dye diffusion, or dye transfer
A head comes in direct contact with the uncoated side of the wax ribbon pushing the inked ribbon to the surface of the paper. Ink is heated and the melted ink transfers to the photo paper surface as a dot pattern (a.k.a. donor/receiver or two-paper system).

INKJET PRINTING
Liquid inks sprayed as dot patterns onto assorted substrates
Inkjet uses cartridge inks both dyes and pigmented dyes. There are three drop-on-demand (DOD) printers: bubblejet, phase change, piezo; and one continuous flow.

- PHASE CHANGE = *Solid to melted to solid, CMYK color stick or wax puck*

Dye in wax applied to paper creates a slightly raised surface which is heat-sensitive. Generally more commercial and prints on many types of substrate.

- PIEZO (a.k.a. MICROPIEZO) = *Liquid or solid/water, solvent or oil based*

Uses droplets of ink squeezed through a nozzle when voltage is applied to a crystal. The crystal pushes a sealed membrane which pushes the dot onto the paper. These are used in fine art and large format images. They print with water- and solvent-based inks, dye and pigment inks.

- THERMAL = *Also known as bubble jet, boil inks, pop glass, jet color to paper*

This is the most common desktop technology. A thermal print head draws ink into a reservoir, where it is heated, pressurized and jetted onto paper surface.

- CONTINUOUS FLOW = *Tight dot pattern that appears continuous*

Such a fine dot pattern is created when jetted, a 300dpi appears to be that of 2000dpi.

PRINT TECHNOLOGY SAMPLES
These samples are only listed from tests between 1996 and 2001, and have been listed only as samples for possible identification assistance.

ELECTROPHOTOGRAPHY = *Dry toner B&W and four-color copies*
    Located in common office supply houses like Kinko's, Office Max, Staples
        Canon CLBP 360PS, CLC550
        Minolta CF900
        Xerox Majestic 5765, 5775, Regal 5790

ELECTROSTATIC PRINTING = *Pigmented toner on dielectric paper and laser printers*
    Large format printers used in commercial applications (three types)
        Image transfer = for outdoor signage
        Dry transfer = for short to long term banner use
        Wet transfer = numerous commercial applications
    Not generally used for fine art
Once printed is often transferred to another substrate (vinyl) via P-S
    Does not use thermal papers, but heat-set ink process
        Ricoh NC5006 = dry electrostatic transfer system
        Ricoh NC8115 = laser electrostatic transfer system
        IBM 3170
        Kodak 1525 ColorEdge

THERMAL TRANSFER PRINTING = *Four-color ribbon printers, dye sublimation*
    Print 40" wide, UV and moisture resistant
    May be used outdoors without cold lamination
    Produces photo realistic images
        Kodak kiosks using Image Magic Paper
        Tektronix Phaser 150, Fuji...

INKJET PRINTING = *Liquid inks sprayed as dot patterns onto assorted substrates*
- PHASE CHANGE = *Solid to melted to solid, CMYK color stick or wax puck*
      Tektronix Phaser 340; LaserMaster DisplayMaker Express.
- PIEZOELECTRIC = *Liquid or solid/water, solvent or oil based*
      LaserMaster Professional = paper, canvas, vinyl
      Xerox, Tektronix, Roland and ColorSpan
      All Epson including 3000, 7000, 9000 series...
- THERMAL = *Also known as bubble jet, boil inks, pop glass, jet color to paper*
      Hewlett Packard HP2000C, HPdeskjet series...
      Lexmark Z51
      Canon printers
- CONTINUOUS FLOW = *Tight dot pattern that appears continuous*
      IRIS 3000 series = large format, variable dot size
      IRIS SmartJet 4012 = medium format, desk size

DIGITAL HEAT TOLERANCE TEST RECOMMENDATIONS

The following digital testing was completed end of 2001 to better understand the heat and laminate tolerances of digital images. Having completed the copier testing in 1996, this updating was necessary to better understand how to handle electronic images as framers. The technology that produces any digital is what determines its heat sensitivities, but not being able to visually identify them is what makes mounting difficult.

The reason an image can be heat-sensitive and not approved for dry mounting at 170-185°F (77 - 85°C) yet still be laminated at higher 225°F (104°C) has to do with the gloss from the fuser rollers. Once a vinyl laminate has been applied over an image with damaged gloss the damage may be covered up and the image may be suitable to frame. But damage can also occur to the inks when laminating. When it has been indicated a digital image does not tolerate laminating it may not be the temperature but rather the mounting adhesive that impacts the resulting lamination. The binders used in the mounting process of some pigment inks are sensitive to the adhesive used to bond the laminate. When heated to laminating temperature they appear to release, float around, and reclump prior to setting. The resulting image appears clumped making fine type fuzzy or difficult to read, and solid color areas very pointillist like a Seurat painting. Though artists may like the abstracted creative image, this is not acceptable for custom framing techniques and signage.

DIGITAL PHOTO HEAT TOLERANCE TEST RECOMMENDATIONS

This test was conducted to better verify the ink, media, printer combinations that better tolerate heat. Since thermal inkjet and bubblejet digital photos are being readily produced at home, proper mounting practices must be determined. A Bienfang 210M-X mechanical press was used to test mount assorted digital photos from a single HPdeskjet 960c printer used as the control. A cross section of digital photo papers were used and since image inks take days to permanently set they were let to sit a week to cure.

Though still inconclusive, most (but not all) of the digital photos tolerated the lowest dry mounting temperature of 150°F. Recommendations will continue to be cold mounting for digitals unless the source is known or a secondary image is available for testing.

Laminating adhesives do not liquefy and reclump the dye inks as they do with pigmented toners from electrophotographic color copies, so laminating is an option for all digitals tested. Both perforated and nonperforated films were tested. Since some ink and paper combinations are more nonporous than others, it is still recommended to use perforated films in the event porosity might be an issue. Numerous digital photos were porous enough to use a nonperforated film. A few images were tested at higher laminating times and temperatures without a film see their tolerance extremes, results are noted.

DIGITAL HEAT TOLERANCE TEST

All tests were conducted using a Bienfang 210M-X mechanical press for 1 minute per temperature setting. YES indicates the sample tolerates heat mounting as the designated temperature at 1 minute. NO indicates there is unacceptable visible damage to the sample at the temperature, even at only 1 minute. A 50/50 notation is based on visual tolerance of image damage. Lighter colors are more tolerable with fair distortion than dark, yet a 50/50 notation specifies there is visual damage at the tested temperature.

Printer	**Type** 1 minute @	150°F	170°F	185°F	200°F	225°F
Minolta CF900	Magnetic roller, Dry pigment toner	yes	no	no	no	no
Xerox 2135	Color Single pass LED	yes	no	no	no	50/50
Xerox 2006	Single pass LED	yes	yes	50/50	50/50	50/50
Canon 360PS	Electrophotography, Dry toner	yes	50/50 light colors OK			no, ink clots
Scitex Spontane	Electrostatic toner copier	yes	no	no	no	ink clots
Kodak ColorEdge	Electrostatic toner copier	yes	no	no	no	ink clots
IBM 3170	Single pass LED	yes	no	no	no	no
Xerox N Series	Black and White Laser	yes	yes	yes	yes	yes
HPdeskjet 960c	Inkjet on Paper	yes	yes	yes	yes	yes
HPdeskjet 960c	**Inkjet on Photo paper**	**no**	no	no	no	no
Kodak Digital Printer	Thermal transfer, Dye transfer, Dye sublimation, Dye diffusion	yes	yes	yes	yes	yes
Tektronix Phaser 140	Liquid Inkjet	yes	yes	yes	yes	yes
Tektronix Phaser 240	Thermal transfer on Film	yes	melts			no
Tektronix Phaser 340	Phase change, Solid ink	yes	melts			no
Tektronix Phaser 440	Thermal transfer	yes	no, transfers to release paper			
Tektronix Phaser 550	Color laser	yes	yes	no	no	no
Tektronix Phaser 750	Color laser	yes	no	no	no	no
Tektronix Phaser 790	Color laser	yes	yes	50/50		yes
Tektronix Phaser 860	Phase change, Solid ink	yes	no, transfers to release paper, soaks through image			
DocuColor 2006	Color laser	yes	yes	no	no	yes
Digital Painter	Inkjet on Canvas	yes	yes	yes	yes	N/A

Upon completion of the above test it was determined additional testing needed to be done with digital photo papers using the same printer to determine the validity of the heat intolerance of the HPdeskjet 960c on Digital Photo Paper. The following chart indicates the results from that test. All tests were completed twice and controlled by using the same Bienfang 210M-X press as above. The 150°F tolerance shows low temperature HA boards and cold methods may be used for mounting with most products. Test before use.

DIGITAL PHOTO HEAT TOLERANCE TEST

Paper	Notes	Time	65°C 150°F	76°C 170°F	85°C 185°F	93°C 200°F	107°C 225°F	
HP Premium Plus Photo Paper		1 min	yes	yes	yes	yes		
HP Premium Plus Photo Paper		5 min					perforate	yes
HP Premium Photo Paper	Hot Press Overlay Foil sticks to inks	1 min	no	no	no	no		
HP Premium Photo Paper		5 min					perforate	yes
HP Photo quality Inkjet Paper Matte		1 min	yes	yes	yes	yes		
HP Photo quality Inkjet Paper Matte		5 min					not perf	yes
HP Photo quality Inkjet Paper Matte		3 min	yes	yes	yes	yes	not perf	yes
HP Photo quality Inkjet Paper Matte		5 min					not perf	yes
HP Photo quality Inkjet Paper Matte	Tested at higher temperature with no laminate	5 min					no laminate	yes
HP Brochure and Flier Paper Matte		1 min	yes	yes	yes	yes		
HP Brochure and Flier Paper Matte		3 min	yes	yes	yes	yes		
HP Brochure and Flier Paper Matte		5 min					not perf	yes
Canon Glossy Photo Paper	Do not reposition laminate, inks peel off easily	1 min	yes	yes	yes	yes		
Canon Glossy Photo Paper		5 min					not perf	yes
Kodak Premium Picture Paper Gloss	Release paper only, no Hot Press Overlay Foil	1 min	yes	yes	no	no		
Kodak Premium Picture Paper Gloss	With Hot Press Overlay Foil	1 min	yes	yes	yes	yes		
Kodak Premium Picture Paper Gloss	Temperature tested no laminate	5 min						no
Kodak Premium Picture Paper Gloss	Release paper only, no Hot Press overlay foil	5 min					perforate	yes
Jetprint Photo Paper	All reacts to heat	1 min	no	no	no	no		no
Epson Photo Paper Glossy	Orange peel texture on base unprinted paper	1 min	yes	yes	yes	yes		
Epson Photo Paper Glossy	Do not reposition laminate, inks peel off	5 min					not perf	yes
HP Multi Purpose Inkjet Paper		1 min	yes	yes	yes	yes		
HP Multi Purpose Inkjet Paper	Heat bleaches inks at highest temperatures	5 min					no laminate	no
HP Multi Purpose Inkjet Paper		5 min					not perf	yes

COPIER HEAT TOLERANCES

This test, completed in 1996, and was set up to establish the duplication accuracy, heat and lamination tolerances for standard electrophotographic toner color copiers used commercially at the time. All images were submitted to operators for duplication on stock papers and the most accurate duplicate of the original was requested. In addition a set of control images was printed on Strathmore 100% cotton rag paper for each.

Results of this test show that some copies are safe to mount at low temperatures. *Speed*Mount HA boards by Bainbridge which mount at 150°F could safely dry mount the copies highlighted in bold text. If the copy origin is unknown then pressure-sensitive applications are suggested. It appears that electrophotographic images may tolerate heat while electrostatic technology remains more heat sensitive.

Copier	Temp	Results
Xerox 5765	**160°F**	**Safe to mount at low temperature**
Majestic	190°F	Higher temps cause surface scuffing, mottles toner
electrophotographic	220°F	Extreme pointillism, No laminates
Xerox 5775	**160°F**	**Safe to mount at lower temperatures**
	190°F	Slight toner reaction
electrophotographic	220°F	Slight pointillism, clotted inks into large dots
Xerox 5790	160°F	Slight toner reaction
Regal	190°F	Visible gloss mottling, like photo scuffing damage
electrophotographic	220°F	Mottled gloss, 50% pointillism, do not Laminate
Ricoh 5006	160°F	Slight visual gloss removal
	190°F	Major toner gloss damage
electrostatic	220°F	Major curdling and color fading, do not laminate
Canon 350	160°F	Almost undetectable change in toner gloss
	190°F	
electrophotographic	220°F	Fuzzy, pointillism 25% worse than original
Canon CJ17	**160°F**	**No difference in color image at any temperature**
	190°F	**or with any laminate overlay, safe to mount**
electrophotographic	220°F	No difference in inks, OK to laminate
Kodak 1525	160°F	Gloss is removed by heat
ColorEdge	190°F	
electrostatic	220°F	Significant damage, do not laminate

In their infancy, toner copiers were rather light fugitive and colors were sometimes altered by the application of heat or laminate. None of the copies in this test exhibited light fading after 3 months of 24 hour a day exposure. Though still valuable, this test completed in 1996 is somewhat outdated. In house testing must still be done to verify tolerances.

COMPARISON CHART

In the age of preservation framing it has become increasingly apparent that numerous items have a tendency to deteriorate over time. Part of this test was to determine the best method for duplicating an image, newspaper clipping or photograph (in 1996) so it could be framed without fear of immediate or long term damage and fading.

These copiers were tested for accuracy of color, lightfastness, crispness, duplication of the original, and heat tolerance. The numbers in the chart rate the seven tested printers from 1 to 7, 1 being the first choice to use for customer replication. The chart shows how individual copiers rated over-all in the tests involving the copying of a Polaroid photograph; B/W vintage studio portrait yellowed with 30 years age, copied in both color mode and B/W mode; a newspaper article with color photo; a textured gray flannel paper certificate with black lettering; an aged 1947 marriage license; and a promotional Otsuka postcard.

Replication Accuracy	Xerox 5765	Xerox 5775	Xerox 5790	Ricoh 5006	Canon 350	Canon CJ17	Kodak 1525
General Copies	3	1	2	6	4	7	5
Polaroid Photograph	2	1	3	4	6	7	5
B / W Photo 1960 Portrait							
Printed Full Color Mode	2	1	-	3	4	5	-
Printed B/W Mode	-	5	3	4	2	-	1
Newspaper Article	2	1	3	4	5	-	-
Certificate (grey flannel paper)	7	5	1	2	3	6	4
Document (1947 marriage license)	2	3	1	5	6	4	7
Color Lithograph Postcard	1	3	2	4	5	-	-
RECOMMENDATIONS							
Dry Mounting	5 50/50	2 yes	4 yes	7 no	3 yes	1 yes	6 50/50
Laminates*	no	yes	50/50	no	50/50	yes	no

*The laminate was a standard matte heat-set vinyl laminate designed for framing, mounted at 220°F for 5 minutes. Laminate adhesives cause the toner to release, migrate and reclump prior to resetting, losing detail and quality. Heat is not recommended, so if lamination is required it must be a cold laminate application. Xerox and Canon are electrophotographic and seemed to tolerate more heat while the Ricoh is electrostatic and more sensitive to heat and laminates.

WHAT IS ART?
The creation of a framing design is a matter of problem solving, consisting of five stages: definition, creativity, analysis, production and clarification. All designs begin with a definition of what is required. Before a photograph can be framed it must first be identified as a color RC, B/W fibre-base, Ilfochrome Classic, or digital. This helps determine what supplies are required and how it should be mounted.

The argument over the validity of art is ongoing. Only a few decades ago the photograph was not considered fine art. Calligraphy continues the fight and now the world of digitals has joined in the struggle for recognition and acceptance.

GICLEES
In the world of digitals images, the giclee has marched onto the scene and established itself as a viable entity in the fine art world. It has achieved status and acceptability as a technical process like the serigraph, monoprint and lithograph. Although a giclee will tolerate dry mounting temperatures just like the above limited editions, it should not be mounted. The question is not, *can it be dry mounted,* but *should it be mounted at all?*

A giclee is a common term used for fine art digital prints, particularly ones printed with continuous flow printers, such as the innovative IRIS. There are two levels of giclee currently being printed: fine art giclee, and decor-quality giclee. Both are inkjet prints but thanks to the Giclee Printers Association (GPA) there have been a set of acceptable standards to keep them separated. Fine art giclees are usually signed and may be numbered and should therefore be framed to preservation standards.

COMPUTER GENERATED (DIGITAL) ART
Unlike the giclee which is a digital rendition of an existing image, computer generated or digital art is created solely within the computer. The printed image becomes the completed project or artwork. Since the 1960s, digital art has fought for acceptance. Just as the photograph has had to prove it was the photographer and not the camera that created the image, the computer artist must prove it is the skilled computer artist that is capable of programming the computer generated art. The camera and the computer are merely the tools used by the artist like a brush or tubes of paint. Successful digital art or computer enhanced images require the skills and talent of a computer artist to manipulate and create the final image. The same computer programs and tools placed in the hands of a nonartist will never produce the unity of design for a successful piece of art.

Though acceptance runs from belief digital art will replace all other mediums to museums who will not purchase digitals for their collections, digital art is now an accepted form. Is taught at respected art schools and exhibited in major museums. Regardless of whether produced on a canvas, through a lens, or on a disk...it **is** art.

BIBLIOGRAPHY

3M, "Mounting, Matting and Framing with 3M Products."
    St. Paul, MN: 3M Stationery and Office Supply Division, 1977.

"Adhesives." COMPTON'S ENCYCLOPEDIA. Volume 1, page 43-44.
    Compton's Learning Co., Chicago, IL., 1991.

Clapp, Anne F. CURATORIAL CARE OF WORKS OF ART ON PAPER.
    4th Edition, New York: Nick Lyons Books, 1987.

Decor Magazine. FRESHMAN FRAMER: BOOK 3.
    St. Louis, MO: Commerce Publishing, 1987.

Drytac Corporation. GREAT WAYS TO PROFIT FROM LAMINATING.
    Richmond, VA: Drytac Corporation, 1995.

"Electrostatic Photocopying." COMPTON'S LIVING ENCYCLOPEDIA.
    Compton's Learning Company, 1996. Online. (January 1997).

Eastman Kodak. CONSERVATION OF PHOTOGRAPHS. Kodak Publication No.F-40.
    Rochester, New York: Eastman Kodak Company, 1985.

Ilford. TECHNICAL INFORMATION: MOUNTING AND LAMINATING CIBACHROME DISPLAY PRINT MATERIALS AND FILMS. Paramus, New Jersey: Ilford Photo Corporation, 1988.

Hot Press Supplies. A GUIDE TO MOUNTING AND LAMINATING.
    Danbury, CT: Hot Press, 1996.

Keefe, L.E. and Dennis Inch. THE LIFE OF A PHOTOGRAPH.
    Stoneham, MA: Butterworth Publishers, 2nd Edition, 1990.

Kistler, CPF, GCF, Vivian Carli. PICTURE FRAMING. The Library of Professional Picture Framing, Volume 1.
    Fairlawn, OH: Columba Publications, 1986.

Kretzmer, E.R. "Facsimile." COMPTON'S LIVING ENCYCLOPEDIA.
    Conpton's Learning Company, 1996. Online. (January 1996).

Lamb, CPF., Allan. FRAMING PHOTOGRAPHY. The Library of Professional Picture Framing, Volume 6.
    Akron, OH: Columba Publishing Co., 1996.

"Photocopying." CONSICE COLUMBIA ELECTRONIC ENCYCLOPEDIA.
    Columbia University Press, 1994. Online. (January 1996).

Seal Products. MOUNTING, LAMINATING AND TEXTURIZING.
    Naugatuck, CT: Seal Products Inc., 1986.

"Thermal Printing." CONSICE COLUMBIA ELECTRONIC ENCYCLOPEDIA.
    Columbia University Press, 1994. Online. (January 1997).

"Types of Printing." COMPTON'S LIVING ENCYCLOPEDIA.
    Compton's Learning Company, 1996. Online. (November 1996).

Wilhelm, H. THE PERMANENCE AND CARE OF COLOR PHOTOGRAPHS.
    Grinnell, Iowa: Preservation Publishing Co., 1993.

INDEX

A
3M Spray 77......................46
3M Vac-U-Mount..............46

A
Abitibi board........................8
accordian support...............45
Acid-Free Mounting Film..13
acids....................................15
acrylic.................................12
adhesive
    removing.......................23
    solvent-based............. ..43
    water-based..................43
adhesives
    composition...................13
    heat-activated..........11, 12
    pH.................................15
    porosity.........................14
    pressure-sensitive..........11
    spray..............................11
    type of bond..................12
    wet..........................12, 23
adjusting pressure...............71
advertising............................9
air bubbles.............39, 54, 102
alcohol cleaning.................42
ANSI..................................54
archival photo boards.........19
average temperature...........34

B
Bainbridge 2U....................21
Bainbridge SpeedMount....57
Beaverboard......................18
Bestine thinner.............22, 31
Bienfang 210M-X............. 29
blueprints......................42, 57
bond....................................13
bond failure...........19, 22, 43
bowing...............................91
brass rubbings..............57, 99
brush application............... 39
brush strokes...................131
bubblejet.....................61, 64
bumps...............................25

C
C-35 PMA Applicator........49
Canon....................146-151
canvas transfer..........119-126
    brush strokes..............131
    canvas.........................123
    copyright....................129
    fibre-base....................128
    paper..........................130
    RC photo....................124
certificates...............5, 52, 139
Cibachrome..... ..5, 56, 62, 63
............101, 120, 132, 146
cleaning........................28, 30
cold face mounting............59
cold laminating..................59
cold mounting....................58
cold pressing....................141
cold vacuum frame............38
............39, 41, 43, 46, 49
..........................50, 57, 59
color copies............5, 60, 102
color tinting.................53, 87
ColorMount................13, 96
Commercial 210M-X........
                         29
composition.......................13
computer art...............80, 145
conservation.........5, 6, 15, 17
............18, 19, 23, 38, 100
.....101, 105, 124, 134, 135
cooling...............................27
copiers................................61
copyright..........................119
corrugated boards..............18
countermount...............17, 40
creative applications.......6, 10
    canvas transfers..........119
    shadow box..................92
    wrapped mats..............88
    wrinkled paper.............90
CrescentAchival Photo...... 19
cyanotype..........................42

D
daily routine................28, 30

D
decorative art........................6
decorative paper.................90
diaphragm.........30, 34, 35, 59
digitals...............6, 57, 62, 142
    photos............61, 87, 106
    samples........................146
    testing...................147-151
digital technologies
    dye sublimation.....64, 148
    continuous flow.............66
    electrophotography.......62
    electrostatic............63, 150
    inkjet......................64, 149
    phase change.........65, 148
    piezo......................65, 146
    thermal transfer.......61, 63
dirt.....................................25
disclaimer.............................6
dry mount adhesives..........13
dry mounting....23, 33, 36, 69
dry toner..........................102

E
edge strips........................101
elements of mounting
    moisture........................34
    pressure.........................34
    temperature..................33
    time..............................32
encapsulation.........42, 59, 107
Epson........................146-151

F
fabric...................6, 12, 52, 93
    heat-activated................19
    neutral pH.....................19
fabric wrapped mat............92
face mounting....................59
facsimile.............................61
FACTS..............................54
fibre-base photo........104, 124
fiber expansion..................40
fiber saturation...................87
Filmoplast-90...................135
flat mount......................... 54

flattening images
   cold..................................60
   hot....................................59
   humidifiers..................140
float mount...................53, 82
floated art...........................41
Flobond..............................13
flush mount..................54, 83
foam boards......................17
foam plastic......106, 109, 114
fugitive..............................19
Fusion 4000......................13

G
Gatorfoam...................18, 62
ghosting.................17, 53, 86
giclee'........................62, 145
glass..................................31
Grumbacher....................127
Gudy 870..........................50

H
hardbed press............20, 35
..........................36, 73, 77, 84
heat-activated foam....15, 144
heat-seal.........................111
heat-sensitive............61, 101
heat tolerances................102
heavy-duty foam boards.....17
heirloom.............................5
Hewlett-Packard.......146, 148
Hot Press.............100, 139, 147
   Overlay Foil..........100, 148
hot pressing....................141
hot vacuum press.......66, 74
..............................77, 98, 118
humidifiers......................140

I
IBM..........................146-151
Ilfochrome Classic.........5, 57
..................63, 101, 132, 146
impact printers..................61
inkjet..................5, 60, 65, 108

continuous flow.............65
   phase change.................65
   piezo..............................65
   thermal..........................65
investment.........8, 10, 38, 137
IRIS............................65, 146

J
Japanese hinges................101
Japanese papercuts........41, 48
jigsaw puzzles...................98

K
Kodak.............................135
Kodak ColorEdge.....146-151
Kodak Film Cleaner.........135
Kraft paper..................39, 46

L
laminating.........6, 10, 15, 109
   basic.............................115
   one-step......................118
   oversized.....................119
   photographs................120
   suffocation...................121
   two-step......................122
laminating creativity............9
large format printers..........146
laser prints......................5, 63
Lexmark..........................146
Library of Congress...........19
................................42, 56
lightfastness....................102
lighting............................25
limited edition....................5
longevity..........................42

M
maintenance................28, 30
marketing...............9, 10, 106
mat boards........................17
Masonite.....................18, 29
McDonalds Acrylic Gel...127
MDF......................16, 44, 62

mechanical press..........20, 21
...................27, 33, 35, 36, 66
..............................68, 77, 84, 97, 117
Melinex.............................42
MightyCore.......................18
Minolta............................146
montage............................81
mounting
   dry.................................69
   elements of....................32
   pressure-sensitive..........49
   spray..............................43
   suggested methods.......136
   wet.................................37
mounting techniques......6, 10
   cold..........................57, 95
   float.............41, 54, 86
   flush.......................54, 87
   montage.................49, 85
   multiple bite..................88
   one-sided......................89
   oversize.............41, 83, 97
   plain........................53, 82
   premounted.............41, 84
   translucents..................89
   trimmed to size.............82
   two-sided......................90
   silks...............................91
mounting creativity......88, 92
MT5............................13, 19
multiple bites.........21, 68, 84
Mylar.......................20, 101
Mylar-D...........................42

N
Neschen Gudy 870.............50
newsprint..........................90
nonimpact printers............60
nonporous..........11, 13, 14, 15
...............18, 41, 45, 54, 76, 97
..............101, 109, 117, 131, 135
nonpositionable.................55
nonreversible....................15
Nucor...............................18

O

one-step
  laminating...................118
  shadow box....................98
orange peel.....16, 95, 96, 101
original art............................5
Overlay Foil/Film...............100
overlay foam..106, 109, 1104
oversize.....................97, 115
  posters..........................41
  photos..........................101

P

papercut art.........................41
paper grain..........................95
papyrus.............................101
parchment........................5, 10
payoff..................................10
PEC 12.......................19, 100
Pelon...................................40
Perfect Mount Film.............51
perforator...................114, 116
permanence........................43
permanent adhesive............13
pH.......................................15
photo boards......................19
photocopies.................60, 61
photograph.......20, 22, 55, 62
.........82, 96, 99 109, 116
  cleaning..........................19
  heat damage..................57
piercing tool......................106
pinholes..............................25
platen...........................24, 31
platen cleaning...................28
Plexiglas......................18, 63
PMA............................34, 50
Polaroid..............................62
pollutants............................31
polyester............20, 56, 58, 6
...................63, 101, 106
plyester encapsulation........59
polyester films..........111, 132
polyflute..............................42
polypropylene....................42
porosity........................14, 43
porous............1, 13, 14, 15, 21
.............24, 45, 52, 54, 56
................78, 79, 80, 81, 84
...............85, 86, 87, 97, 99
................100, 115, 116, 135
Positionable Mounting
  Adhesive (PMA)...........49
posters........................6, 133

preadhesive boards............15
predry..............26, 33, 36, 73
premounting.....................125
press maintanence........28, 30
pressure.........59, 66, 71, 107
  adjustments...................75
  gauge............................74
  shimming......................76
pressure-sensitive.........33, 36
  boards...........................50
  films..............................50
  mounting......................49
  tapes.............................51
pricing............8, 15, 106, 114
printing
  bubblejet......................62
  electrostatic.................63
  inkjet......................62, 64
  laser.............................63
profit..............8, 9, 10, 127
psi..............................35, 78
pure film adhesives............13
puzzles...............................98
PVA......................11, 12, 41

R

raw canvas.......................125
RC photograph.......6, 24, 41
..............45, 56, 62, 96, 99
release board..............21, 122
release papers....................20
  Mylar.....................24, 101
removable.........13, 15, 23, 24
repositionable..............38, 54
reversible....5, 23, 45, 63, 105
rice paper...........................85
Ricoh........................146, 150
rigid board stiffener..........122
Roland.............................146
roller laminators..........49, 58
........................109, 142

S

sales........................... 7, 9
samplers...............................7
scuffing.............................100
Seal Graphics...................102
self-shaping.................46, 53
shadow box.........................92
sheer fabric........................87
shims..................................72
silk......................................87
silver gelatin.....................100

Sintra..................................18
softbed press................36, 88
solvents............12, 22, 24, 31
SpeedMount.......57, 144, 150
sponge pad.......20, 21, 27, 29
sprays.........12, 23, 34, 42, 45
Spraytex
  Good Glue Spray..........45
starch..........................11, 12
static mounting.................67
statistics...............................7
steam....................35, 73, 76
storage...............................26
substrate.......................16-19
suction sealing...................86
Super 77.............................45
Super Unimount.................13
Sure Mount Spray..............45
surface laminating....105, 113
surface tacking...........67, 114

T

tacking
  multiple bite.................73
  surface..........................71
  z-method......................72
tacking iron.......................67
Tektronix..................146-149
temperature......33, 37, 42, 48
..................57, 65, 107, 108
tapes..................................23
tear strength......................43
thermal wax transfer.....61-64
thermographic.........5, 58, 60
thermography
  photocopying................61
  printing.........................61
  facsimile.......................61
thermoplastic....11, 12, 22, 23
thermosetting.....................11
thickness............................16
time.................32, 37, 42, 48
................57, 65, 107, 108
  bond time................42, 43
  draw time..................... 33
  dwell time.........32, 37, 65
  open time....32, 38, 43, 46
  tack time.......................32
TM1- TM4.........................13
toluene...............................22
Trimount............................13
TTPM...............32, 37, 42, 43
.................48, 57, 65, 66
two-step............102, 117, 122

U

unhinging............................29
un-du................................22
UnSeal........................22, 31

V

VacuGlue 300..................123
vacuum press..........33, 35, 36
vellum...........................5, 101
vinyl........................106, 116
vinyl films........102, 112, 132

W

warp..............................26, 91
warping..................16, 17, 18
....................26, 35, 40, 91
watercolor paper...............125
water soluble......................12
weighting......................40, 66
weights..............................43
wet glue......12, 32, 35, 37, 39
wet mounting.....................37
Winsor/Newton
   Acrylic Mediums........127
   Impasto Gel.................127

X

xerography.........................62
Xerox.........................86, 147

Z

Z-method........68, 96, 99, 114

Trademarks ™/Registered Trademarks ®/Copyrights ©

All copyrights, trademarks and service marks are the property of their respective owners.

ORDER FORM
THE MOUNTING AND LAMINATING HANDBOOK, SECOND EDITION
By Chris A. Paschke, CPF GCF

The original 1997 edition was the most complete and comprehensive manual of its kind written for framers and photographers, which was designed as a technical reference book geared to both beginners and seasoned veterans. It is a complete source for basic techniques, hands-on procedures, applications, and special tips. It has filled the mounting void and been the premier reference handbook on mounting and laminating for framers to date. This new 2002 **Second Edition** keeps all of the basics while adding sixteen NEW pages, including digital explanations for electrophotographic, electrostatic, thermal and inkjet technologies. It also features mounting suggestions, over all tips, and heat tolerance test results for commonly framed digital art and digital photographs.

Learning the basics including premounting, multiple bites, color tinting, and fabric wrapping, allows for applying them to individual framing jobs rather than specific projects...while canvas transferring brings in the profit dollars. By better understanding the world of digitals, their production and niche, the professional custom framer is better geared to framing in the 21st century.

This 2002 release from **Designs Ink PUBLISHING**, is a laminated, soft cover, 176 page edition, 7 x 9", retailing for $19.95. Book is written in essay format, but includes bullet point step-by-step, line art, diagrams, special tips, detailed reference, test results, appendix and bibliography. This new edition also includes an expanded contents and index for easier reference.

For additional information or to mail order please contact:
Designs Ink PUBLISHING
#183-785 Tucker Road . Suite G . Tehachapi, CA 93561
661.821.2188 . Fax: 661.821.2180 . DesignsInk@aol.com www.designsinkart.com

---

*Please print clearly and make checks or money order payable to **Designs Ink** PUBLISHING, in US dollars. www.designsinkart.com*
Allow 2 weeks for delivery. Foreign orders remit double the book cost to cover freight costs.
Send orders with payment enclosed to Designs Ink . #183-785 Tucker Road . Suite G . Tehachapi, CA 93561

Enclosed is $19.95 per book, plus $4.50 for shipping and handling per copy of
The Mounting And Laminating Handbook

NAME _____
Company _____
Address _____
_____
City/State/Zip _____
Phone/Fax _____

Quantity   Price   Total
_____   @ $19.95 = _____
CA residents add 7¼ % Tax _____
Plus $4.50 each book _____

Grand Total _____

ORDER FORM
CREATIVE MOUNTING, WRAPPING AND LAMINATING
By Chris A. Paschke, CPF GCF

This is the must have companion to *The Mounting and Laminating Handbook*. It is the manual written for innovative and creative picture framers and photographers who want to achieve the most from their dry mounting equipment and design skills. Designed as a technical reference book geared to both beginners and seasoned veterans, it is a complete source for unusual and traditional techniques, materials, equipment, and tips.

Written in easy-to-follow chapter format, it covers materials and equipment for creative applications for mounting of tiered mats; wrapped and embossed mats; shadow boxes; and laminating variations of all kinds from resurfacing and retexturing laminates, to faux glass etching. The image transferring chapter is currently the most complete in the industry for transferring to canvas, linen and watercolor papers.

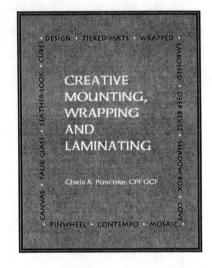

Once an understanding of the basics of mounting and laminating are mastered then applying them to decorative papers, fabrics and foam boards allows for the ultimate in custom framing and profits in design creativity.

This 2000 release from **Designs Ink PUBLISHING**, is a laminated, soft cover, 176 page edition, 7 x 9", retailing for $19.95. Book is written in essay format, but includes bullet point step-by-step, line art, diagrams, tips, glossary, materials sources, contents and appendix, plus complete index and bibliography.

For additional information or to mail order please contact:
Designs Ink PUBLISHING
#183-785 Tucker Road . Suite G . Tehachapi, CA 93561
661.821.2188 . Fax: 661.821.2180 . DesignsInk@aol.com www.designsinkart.com

---

*Please print clearly and make checks or money order payable to **Designs Ink** PUBLISHING, in US dollars. www.designsinkart.com*
Allow 2 weeks for delivery. Foreign orders remit double the book cost to cover freight costs.
Send orders with payment enclosed to Designs Ink . #183-785 Tucker Road . Suite G . Tehachapi, CA 93561

Enclosed is $19.95 per book, plus $4.50 for shipping and handling per copy of
Creative Mounting, Wrapping and Laminating

	Quantity	Price	Total
NAME _____	_____	@ **$19.95** =	_____
Company _____		CA residents add 7¼ % Tax	_____
Address _____		Plus $4.50 each book	_____
City/State/Zip _____			
Phone/Fax _____		**Grand Total**	_____

COMING

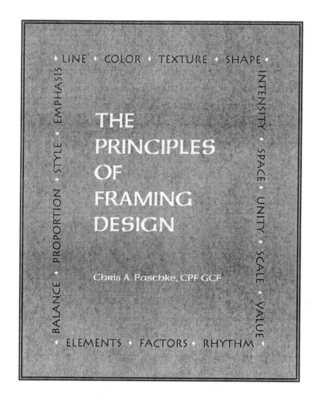

THE BEST IS WORTH WAITING FOR